Springer Finance

Springer Finance

Springer Finance is a programme of books addressing students, academics and practitioners working on increasingly technical approaches to the analysis of financial markets. It aims to cover a variety of topics, not only mathematical finance but foreign exchanges, term structure, risk management, portfolio theory, equity derivatives, and financial economics.

Ammann M., Credit Risk Valuation: Methods, Models, and Application (2001)
Back K., A Course in Derivative Securities: Introduction to Theory and Computation (2005)
Barucci E., Financial Markets Theory. Equilibrium, Efficiency and Information (2003)
Bielecki T.R. and Rutkowski M., Credit Risk: Modeling, Valuation and Hedging (2002)
Bingham N.H. and Kiesel R., Risk-Neutral Valuation: Pricing and Hedging of Financial Derivatives (1998, 2nd ed. 2004)
Brigo D. and Mercurio F., Interest Rate Models: Theory and Practice (2001, 2nd ed. 2006)
Buff R., Uncertain Volatility Models – Theory and Application (2002)
Carmona R.A. and Tehranchi M.R., Interest Rate Models: An Infinite Dimensional Stochastic Analysis Perspective (2006)
Dana R.-A. and Jeanblanc M., Financial Markets in Continuous Time (2003)
Deboeck G. and Kohonen T. (Editors), Visual Explorations in Finance with Self-Organizing Maps (1998)
Delbaen F. and Schachermayer W., The Mathematics of Arbitrage (2005)
Elliott R.J. and Kopp P.E., Mathematics of Financial Markets (1999, 2nd ed. 2005)
Fengler M.R., Semiparametric Modeling of Implied Volatility (2005)
Filipović D., Term-Structure Models (2009)
Fusai G. and Roncoroni A., Implementing Models in Quantitative Finance (2008)
Geman H., Madan D., Pliska S.R. and Vorst T. (Editors), Mathematical Finance – Bachelier Congress 2000 (2001)
Gundlach M. and Lehrbass F. (Editors), CreditRisk$^+$ in the Banking Industry (2004)
Jeanblanc M., Yor M., Chesney M., Mathematical Methods for Financial Markets (2009)
Jondeau E., Financial Modeling Under Non-Gaussian Distributions (2007)
Kabanov Y.A. and Safarian M., Markets with Transaction Costs (2009)
Kellerhals B.P., Asset Pricing (2004)
Külpmann M., Irrational Exuberance Reconsidered (2004)
Kwok Y.-K., Mathematical Models of Financial Derivatives (1998, 2nd ed. 2008)
Malliavin P. and Thalmaier A., Stochastic Calculus of Variations in Mathematical Finance (2005)
Meucci A., Risk and Asset Allocation (2005, corr. 2nd printing 2007, Softcover 2009)
Pelsser A., Efficient Methods for Valuing Interest Rate Derivatives (2000)
Platen E. and Heath D., A Benchmark Approach to Quantitative Finance (2006, corr. printing 2010)
Prigent J.-L., Weak Convergence of Financial Markets (2003)
Schmid B., Credit Risk Pricing Models (2004)
Shreve S.E., Stochastic Calculus for Finance I (2004)
Shreve S.E., Stochastic Calculus for Finance II (2004)
Yor M., Exponential Functionals of Brownian Motion and Related Processes (2001)
Zagst R., Interest-Rate Management (2002)
Zhu Y.-L., Wu X., Chern I.-L., Derivative Securities and Difference Methods (2004)
Ziegler A., Incomplete Information and Heterogeneous Beliefs in Continuous-time Finance (2003)
Ziegler A., A Game Theory Analysis of Options (2004)

Preface

It was the end of 2005 when our employer, a major European Investment Bank, gave our team the mandate to compute in an accurate way the counterparty credit exposure arising from exotic derivatives traded by the firm. As often happens, exposure of products such as, for example, exotic interest-rate, or credit derivatives were modelled under conservative assumptions and credit officers were struggling to assess the real risk. We started with a few models written on spreadsheets, tailored to very specific instruments, and soon it became clear that a more systematic approach was needed. So we wrote some tools that could be used for some classes of relatively simple products. A couple of years later we are now in the process of building a system that will be used to trade and hedge counterparty credit exposure in an accurate way, for all types of derivative products in all asset classes. We had to overcome problems ranging from modelling in a consistent manner different products booked in different systems and building the appropriate architecture that would allow the computation and pricing of credit exposure for all types of products, to finding the appropriate management structure across Business, Risk, and IT divisions of the firm.

In this book we describe some of our experience in modelling counterparty credit exposure, computing credit valuation adjustments, determining appropriate hedges, and building a reliable system.

What do we mean by all of this? Counterparty credit exposure is the amount a company could potentially lose in the event of one of its counterparties defaulting. At a general level, computing credit exposure entails simulating in different scenarios and at different times in the future, prices of transactions, and then using one of several statistical quantities to characterise the price distributions that has been generated. Typical statistics used in practice are (i) the mean, (ii) a high-level quantile such as the 97.5% or 99%, usually called *Potential Future Exposure* (PFE), and (iii) the mean of the positive part of the distribution, usually referred to as *Expected Positive Exposure* (EPE).

With these measures and default probability information or counterparty CDS premia, it is then possible to price counterparty risk. In the financial industry, the economic value of this risk is generally called *Credit Valuation Adjustment* (CVA).

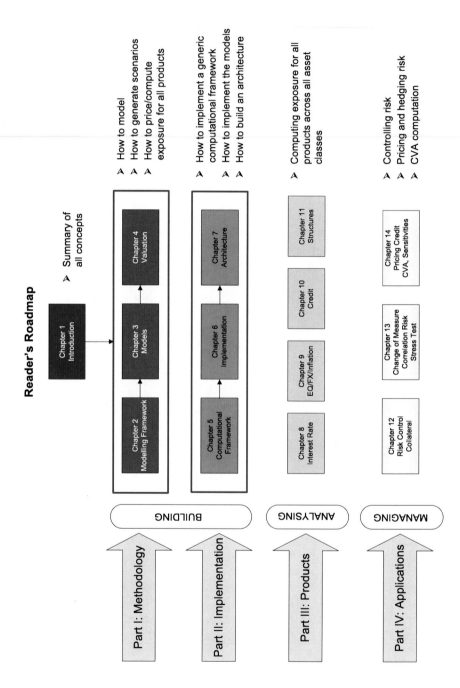

Reader's Roadmap

As we will have occasion to see later in this book, it can be computed as the price of a Credit Default Swap paying the Expected Positive Exposure. Equivalently expressed, CVA is the price of a new type of hybrid product, the so-called *Contingent Credit Default Swap* (C-CDS), which pays the value of the exposure (floored at zero) upon default of the counterparty.

The credit crisis which started in 2007 clearly shows why it is of crucial importance for any company entering a derivative business to (i) measure counterparty exposure, (ii) compute capital requirements, and (iii) hedge counterparty risk. Measuring counterparty exposure is important for setting limits on the amount of business a firm is prepared to do with a given counterparty; hedging it gives a possibility of mitigating it and transferring risk; and from a regulatory perspective there is significant pressure on financial institutions to have the capability of producing accurate risk measures to compute capital. In addition, computing counterparty exposure can also give insights into prices of complex products in potential future scenarios. It seems that what was until recently a Risk Control function attracting relatively limited attention, is now becoming a central activity of all major financial institutions, requiring significant resources from all parties.

Our approach to counterparty credit exposure analysis is quantitative. The focus is on mathematical modelling, simulation techniques using various Monte Carlo approaches, and pricing. In contrast, we are only marginally interested in assessing the quality of counterparties or in analysing historical market data in order possibly to forecast future behaviours of the economy, or in risk and regulatory aspects of the problem. We consider derivative products and complex structures which are usually traded in Investment Banks, and focus on practical aspects of the problems at hand. All models used in our analyses are tested with practical data and real transactions. Given this quantitative focus, we sometimes refer to our work as Credit Quantification.

The book is divided into four parts, (I) Methodology, (II) Architecture and Implementation, (III) Products, and (IV) Hedging and Managing Counterparty Risk. In Part I we present a generic simulation framework, which can be used to compute counterparty exposure for both vanilla and exotic products. We show how the classical Monte Carlo framework, where price distributions are computed by generating thousands of scenarios and by explicitly pricing the product at each point in time and at each scenario, is a special case of our more general framework. The classical Monte Carlo approach works well only for products that can be priced in analytical or quasi-analytical form. It is not practical for products that cannot be priced in closed form and require, for instance, a Monte Carlo or lattice pricing approach. Typical examples are products with callability features or exotic interest-rate transactions. We show how in these cases American Monte Carlo techniques used generally for pricing can also be applied efficiently to compute exposure, as they provide intermediate valuations over time and scenarios.

Part II shows how our simulation framework naturally leads to the implementation of a software architecture and the definition of a programming language that allows the computation of both vanilla and exotic products in a scenario-consistent way. In practice, in a large financial institution one of the main problems in building counterparty exposure systems, is to integrate different products, booked in different systems and priced using libraries written in different languages and with different technologies, in order to compute portfolio exposure across different businesses. We show that our approach leads to an architecture that can integrate other systems in a natural way.

In Part III we consider how to compute exposure for different products. We show how the general techniques and models described in Part I and the architecture described in Part II can be used in practice.

Finally in Part IV things are put together. We consider how to perform risk management and risk control of counterparty exposure on a portfolio basis. We describe different aggregation techniques and a standard set-up that uses collateral to mitigate exposure. We also analyse how to model wrong-way/right-way exposure, where transaction price fluctuations and quality of the counterparty are correlated and we address the problem of changing the reference probability measure after the simulation has been performed. The final chapter is dedicated to pricing counterparty credit exposure and to computing CVA and CVA sensitivities not only to credit spread, but also to market risk factors. The whole book can be seen as a roadmap to achieve this goal.

One note to conclude: in our work we describe and use well-established simulation and pricing techniques. Our goal is not to suggest new or more sophisticated algorithms. It is rather to show how well-known algorithms can be put together and used to compute counterparty credit exposure and which limitations have to be taken into consideration if we want to move from vanilla products to complex exotic transactions.

London, September 2009

Giovanni Cesari
John Aquilina
Niels Charpillon
Zlatko Filipović
Gordon Lee
Ion Manda

Acknowledgements

This book developed from the experience gained during a long-term project within the Investment Bank we work for. As such it would have never been written without the support, advice and encouragement of several people to whom we would like to express our gratitude.

Duncan Rodgers and Myles Wright gave us full support in the development of the ideas described in this book.

Darryll Hendricks and Tom Daula helped us to have a global approach to credit exposure computation and inspired us in the early stages of this project.

From Thomas Hyer and Trevor Chilton we gained a better understanding of the challenges of the American Monte Carlo (AMC) algorithms, while Yuan Gao engineered an early prototype that used AMC not just for pricing but also to estimate credit exposure.

Helmut Glemser kept us on our toes by insisting on explanations (or corrections!) whenever our calculations gave results that puzzled him.

Sincere thanks go to our colleagues Rong Fan and Yi Yuan, who, from across the Atlantic, contributed to the writing and testing of a significant part of our code.

We would like also to thank Mark Davis and Martijn Pistorius for their very useful comments, and Catriona Byrne and her team for their continuous support during the publishing phase of this book.

Thanks are also due to the following people for useful discussions during the course of the project: Richard Adams, Annette Alford, Rowan Alston, Philip Anderson, Ashima Bhalla, Rajinder Basra, Marc Baumslag, Alan Baxter, Lucia Bonilla, Gareth Campbell, Denton Capp, Ben Cassie, Dean Charette, Paul Charles, Dipak Chotai, Jack Chung, Mark Dahl, Valdemar Dallagnol, Gerald Elflein, Jesus Fernandez, Alex Ginzburg, Alasdair Gray, Lionel Guerraz, Stephen Johnston, Jeffrey Lin, Catarina Lopes, Felix Matschke, David Matthews, Peta McRae, Sourav Mishra, Andrew Morgan, Bruno Mugica, Logan Nerio, James Ntuk-Idem, Sarah Peplow, Tom Prangley, David Shieff, Andrew Tseng, and Winnie Zheng.

Disclaimer

The views and opinions expressed in this book are those of the authors and are not those of UBS AG, its subsidiaries or affiliate companies ("UBS"). Accordingly, UBS does not accept any liability over the content of this book or any claims, losses or damages arising from the use of, or reliance on, all or any part of this book.

Nothing in this book is or should be taken as information regarding, or a description of, any operations or business practice of UBS. Similarly, nothing in this book should be taken as information regarding any failure or shortcomings within the business, credit or risk or other control, or assessment procedures within UBS.

About the Authors

Giovanni Cesari is Managing Director at UBS. He has more than 10 years' experience in modelling and pricing counterparty credit exposure. Before moving to finance, Giovanni worked for several years in particle physics and in theoretical computer science. Giovanni holds a *Laurea* in Electrical Engineering from the University of Trieste, a *Perfezionamento* in Physics from the University of Padova, and a Ph.D. from ETH, Zurich.

John Aquilina holds an M.Phil. in Statistical Science from the University of Cambridge and a Ph.D. in Mathematical Finance from the University of Bath. He has worked on modelling counterparty credit exposure at UBS since 2005.

Niels Charpillon holds a *Diplôme d'Ingénieur* from Ecole des Mines, an M.Sc. in Financial Mathematics from Warwick Business School, and a *Licence* in Economics from University of St. Etienne. He joined the counterparty exposure team at UBS in 2006.

Zlatko Filipović started working for UBS in 2005 as a Quantitative Analyst in the counterparty exposure team. Before joining UBS, Zlatko had been working for Mako Global Derivatives, London, as a Financial Engineer. Zlatko obtained a Ph.D. in Quantitative Finance from Imperial College, London, after graduating from the Faculty of Mathematics, University of Belgrade.

Gordon Lee joined the counterparty exposure team at UBS in 2006. Prior to UBS, he was a Senior Associate in quantitative risk and performance analysis at Wilshire Associates. Gordon holds an M.A. in Mathematics from Churchill College, University of Cambridge.

Ion Manda holds a *Diploma de Inginer* from the University of Bucharest and a M.Sc. in Financial Engineering from University of London. He has been working in the credit exposure team at UBS since 2006. Ion has about 10 years' experience as a software engineer.

Contents

Part II   Architecture and Implementation

Part III Products

Part IV Hedging and Managing Counterparty Risk

Part I
Methodology

A Typical Swap Exposure Profile

Chapter 1
Introduction

The aim of this first chapter is to introduce basic notions of counterparty credit exposure, and to motivate with a few simple examples the problems and concepts we will be considering in more detail later in this book.

1.1 Basic Concepts

Consider two parties, A and B say, who enter into an OTC transactions portfolio.[1] This portfolio could consist of products ranging from interest-rate and cross-currency swaps in different currencies with various exotic features, to exotic options on equity, foreign exchange and commodity underlyings. It could also include various types of credit derivatives contracts, such as credit default swaps (CDS) on single names or collateral debt obligations (CDO) tranches in swap form on portfolios of reference entities, or credit indices.[2]

In general a given company, say a financial institution A, will have portfolios with many other counterparties, varying among sovereign entities, corporates, hedge funds, insurance companies (including for examples monolines[3]). It may also happen that the credit quality of the counterparty is not independent of the performance of the transaction entered into, such as what happens for example, when an electricity generating oil-powered plant bets on the price of oil.

Counterparty credit exposure is the amount a company, say A, could potentially lose in the event of one of its counterparties defaulting. It can be computed by simulating in different scenarios and at different times in the future, the price of the

---

[1] An OTC (Over The Counter) transaction is a transaction that is not traded through an exchange.

[2] A typical credit index is for example iTraxx; it is composed of the 125 most liquid CDS names referencing European investment grade credits.

[3] Monoline insurance are companies that guarantee to bond investors the payment of coupon and notional. They insure different type of securities, such as CDO, structured products and municipal bonds. Monolines have been affected in the recent credit crunch, raising counterparty risk issues.

G. Cesari et al., *Modelling, Pricing, and Hedging Counterparty Credit Exposure,*
Springer Finance, DOI 10.1007/978-3-642-04454-0_1,
© Springer-Verlag Berlin Heidelberg 2009

transactions with the given counterparty, and then by using some chosen statistic to characterise the price distributions that have been generated. Typical statistics used in the industry are (i) the mean, (ii) the 97.5% or 99% quantile, called *Potential Future Exposure* (PFE), and (iii) the mean of the positive part of the distribution, referred to as the *Expected Positive Exposure* (EPE). We will also have occasion to speak about less commonly used statistical measures that can be more appropriate for certain products.

As important as measuring counterparty exposure, via PFE or EPE, is the computation of the cost of hedging it, and the capability of having a dynamic hedging strategy, i.e. the computation of exposure sensitivities. In the financial industry the price of hedging is generally called *Credit Valuation Adjustment* (CVA). We will see that there are strong links between EPE and CVA computation.

1.2 Preliminary Examples

Some simple examples will help clarifying these points.

1.2.1 Vanilla Interest-Rate Swap

Consider counterparties A and B who enter into an interest-rate swap where A receives every six months the 6-month Libor rate on a notional of $100 million, while paying to B a fixed amount equal to the par 10-year swap rate on the same notional observed at inception.

This is a typical swap contract with value zero at inception. As time passes and market conditions change, the value of the swap changes accordingly. Thus, if the

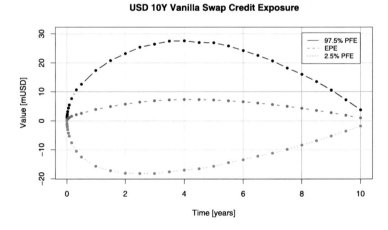

Fig. 1.1 Exposure profile for a typical USD 10-years swap contract, paying fix and receiving floating on a notional of 100 mUSD. The full distribution is shown in Fig. 1.2

swap rate decreases (resp. increases), the transaction will be out of the money (resp. in the money) as seen from A's point of view. Therefore, if B were to default at a point in the life of the trade when swap rates had increased, then A would need to replace in the market—*at higher cost than the fixed amount being paid to B*—the floating cashflows promised and not delivered by B.

To compute the credit exposure for the swap, we would need to estimate the values the swap could take in different market scenarios at points in the future. Figure 1.1 shows the 97.5%, the 2.5% quantiles and the EPE of the swap price distribution, over its entire life, as seen from party A's point of view. A plot like this is usually referred to as the exposure profile. Note that the 2.5% quantile seen from A's perspective, corresponds to the 97.5% quantile seen from B's perspective.

Figure 1.2 shows in the top panel the full price distribution over time. The bottom panel shows three slices of this distribution at three different points in time.

For this example, the 97.5% PFE quantile is a function that starts at zero, peaks at around the 4-year point and then decreases to zero. First, by definition, the fixed payment in the trade is the fair value for the swap, and this must therefore have value (and hence exposure level) identically equal to zero at inception. Similarly, towards the end of the transaction, when all payments but one due under the swap have been paid, the exposure remaining is that from only a single coupon exchange. This explains what happens at the right end of the profile. At intermediate times, the shape of the profile is the result of opposing effects. On the one hand, as the interest rates underlying the swap diffuse, there is more variability in the realised Libor rates, potentially leading to higher exposure. On the other hand, as time evolves there are fewer payments remaining under the swap, and this mitigates the effect of diffusing rates.

The profile therefore tells us that with 97.5% probability, the loss of A due to default of B will not exceed roughly $28 million. Of course, this estimate is based on market information at inception of the swap, and would change if it were to be repeated at a different time.

1.2.2 Cancellable Swap

We can make our example slightly more interesting. It is common for swaps to trade with an additional callability feature, whereby one counterparty would have the option, at certain times in the life of the swap, to cancel ("call") the transaction for a fixed fee (which may be zero).[4]

Suppose that party A, from whose perspective we look at exposure, also holds the option to cancel the trade; one says that A is *long callability*. Assuming A behaves

---

[4]We define "cancellable swap" a swap which has an embedded option to terminate it at zero cost (or at a given predetermined fee). Sometimes these swaps are also called "callable". We use the term callable swap in a more generic way, considering the possibility that the swap is "called" into a new product. In this sense a cancellable swap is a simple example of a callable swap.

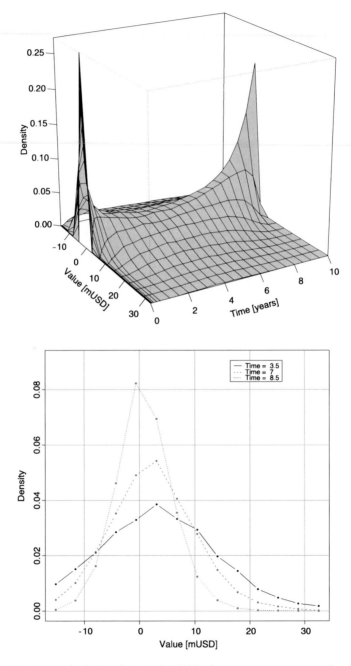

Fig. 1.2 Future value distribution for a typical USD 10-years swap contract, paying fix and receiving floating. The PFE and EPE are shown in Fig. 1.1

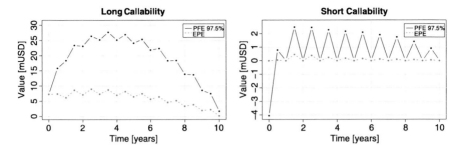

Fig. 1.3 Exposure of a typical cancellable 10-years swap, paying yearly the fair swap rate fixed at inception and receiving semi-annually the 6-month Libor rate on a notional of 100 mUSD. On the left (resp. right), the exposure represents long (resp. short) optionality to cancel the swap every year. The value of the swap at time-zero corresponds to the value of the option, which is assumed to be paid up-front by the counterparty

rationally, it would never decide to walk away from the swap in those scenarios where the swap has a high value (because the swap rate has increased and future receivables are worth more than at inception). This means that having the option to cancel, at zero cost, should not affect materially A's exposure. On the other hand, suppose A is *short callability*, meaning that it is B who has the option to walk away from the swap. Rational behaviour on B's part implies that B would cancel the swap when they are making a loss on the transaction, which is exactly when A would be in the money. Thus we would expect that with A being short callability, A's PFE (and EPE) to B is reduced to zero at each date where B has the option to cancel the transaction.

Figure 1.3 shows all this happening. In the left panel, we see that with A having the option to cancel the trade, the PFE profile is similar to that of a vanilla swap, with the exception of the time-zero level, which equates to the value of the cancellation option. On the right we see that A's exposure is reduced to zero at dates where B can cancel; on remaining dates, the PFE is reduced to that arising from coupons due until the next allowed cancellation date. Note that the time-zero point is not zero but negative from A's point of view, since it is B who holds the option in this case. Note that in practice the value of the option is often embedded in the fixed coupon of the swap, which has then value zero at inception.

From the computational point of view, there is a fundamental difference between the vanilla swap example of the previous section and the cancellable swap we have just described. A vanilla swap can in fact be priced analytically and in a model independent way, and therefore, as we will see, exposure could be computed in a classical Monte Carlo framework, where scenarios are generated and then products are priced at each scenario and each time step. On the other hand, a cancellable swap is *priced* using a lattice or Monte Carlo simulation, making therefore impractical the computation of credit exposure itself by Monte Carlo simulation.[5] This would entail in fact a Monte Carlo of Monte Carlo approach (with nested simulations), where one set of simulation is used for scenarios and one set, at each time step and scenario, for pricing. We will analyse this aspect in more details in the next chapters.

1.2.3 Managing Credit Risk—Collateral, Credit Default Swap

When structuring a new transaction (or portfolio of transactions), one of the criteria is the amount of acceptable credit exposure. This will depend on several factors including risk appetite and quality of the counterparty. The most common way to reduce counterparty exposure is to set up a collateral agreement, whereby the client is required to deposit collateral into a separate account at regular time intervals. A collateral agreement between counterparties can take one of several forms. For instance, it can be in the form of cash or securities, can be called daily or at other regular intervals, and there can be thresholds and minimum transfer amounts. In addition, since the point of holding collateral is to be able to liquidate it in case of the counterparty defaulting, market liquidity plays an important role in determining the amount of collateral needed. When collateral agreements are in place, therefore, credit exposure computation has to take into account features of that agreement together with the dynamics of the trade itself, in order to compute so-called *close-out risk*. Close-out risk measures the amount by which the value of a transaction could change during the period from when the counterparty is deemed to have defaulted, until the collateral has been liquidated and used to fund, at current market conditions, the replacement of the defaulted counterparty in the transaction. In general this computation should also include change in value of the collateral, possibly taking into account the correlation between collateral and transaction value.

A further possibility for A to manage counterparty credit exposure to B is to buy Credit Default Swap (CDS) protection on B from another counterparty C. The transaction between A and C is typically fully collateralised. This will transfer the risk of B to C. In case of default of B, the CDS would ensure that C will step in and make good any payments that were originally promised by B, or simply pay the value of the transaction. This should cover the value of the products (e.g. the interest-rate swap we described before) as calculated at the time of B's default. The value of the protection is called Credit Valuation Adjustment (CVA) and in principle should be charged to the client (in our case counterparty B) in order to reflect its credit risk.

For instance, suppose the credit spread of B is 100 bps,[6] the amount to protect $100 million, the trade maturity 10 years. Under these market conditions the price

---

[5]Note that a cancellable swap is the combination of a vanilla swap and a Bermudan option. If the option is European (i.e. the swap can be called only on one date), the cancellable swap can be priced in closed form.

[6]bps: basis points, a hundredth of a percent.

of buying protection on B will be in the order of $10 million.[7] Such protection is sufficient only at the time of calculation, and one would need to compute exposure sensitivities to the underlying factors in order to dynamically hedge the required amount of protection on B, as the future exposure to B will evolve with market conditions.

As we mentioned, the usage of CDS transfers the credit exposure from B to C. So, even if one assumes the exposure to B to be perfectly hedged via the CDS, there will be counterparty exposure to C, which offers protection.[8] Consider again the example above where A buys CDS protection on B on a notional amount of $100 million. Figure 1.4 shows a typical profile for such a transaction, assuming it is un-collateralised. For such a default protection product, the exposure one observes

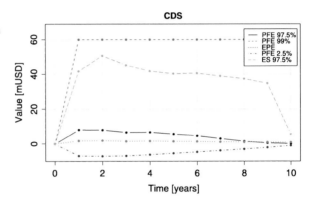

Fig. 1.4 Exposure of a typical credit default swap on a notional of 100 mUSD and spread about 100 bps. PFE (on different quantiles), EPE and ES are shown

results from the effects of (i) movements in the simulated credit spread of B and (ii) defaults. Clearly, the payment triggered by B's default, equal to about $1 - R = 60\%$ of notional, would imply that in a default scenario, A would have an exposure to C of $60 million. R is the recovery rate, i.e. the amount which can be recovered upon default of the counterparty. Now in Fig. 1.4, the PFE profile (which we recall is the 97.5% quantile of the distribution) does not show such high levels of exposure. This must mean that in the simulated scenarios, fewer than 2.5% of the scenarios involve B defaulting. Or in other words that the event of B defaulting is a rare event. To take into account this event one could display higher quantiles of the distribution, say the 99.9% quantile. Alternatively, one can calculate the *Expected Shortfall (ES)* of the distribution, which is simply the expected value of the tail of the distribution (see Chap. 12 for more details); this measure will uncover any large outliers in the distribution (such as the rare event of default of B, and hence large payment by C, in this case). Figure 1.4 displays this quantity, and clearly shows that defaults are indeed occurring even if they are not frequent enough to affect the 97.5% PFE.

---

[7]This is roughly equal to the 10-years duration multiplied by the 10-years spread.

[8]This is one of the reasons why, after the 2007–08 credit events, it is under discussion to use clearing houses when dealing with credit default swaps.

This example shows that with credit products, where events of *small* probability can lead to *large* payments, the PFE might not be the appropriate exposure measure to consider. We will have more to say on this in due course.

1.3 Why Compute Counterparty Credit Exposure?

Counterparty risk is at the root of traditional banking. Historically, the first form of financial instruments were bonds, and their value was mainly driven by the market's view of how creditworthy the issuers of these bonds were. However, today's financial world is much more complex, and the process of estimating counterparty risk much more challenging. While for loans and other traditional products the focus is mainly on estimating the capability of the borrower to repay its obligation, for derivative transactions, estimating accurately the future value of the transaction is as important and challenging as having a view on the ability of the counterparty to honour its obligations.

Accuracy is important because credit exposure models are used for several purposes in financial institutions, such as

  (i) Setting limits on the amount of business allowed with a particular counterparty.
 (ii) Dynamic hedging of counterparty risk, by buying credit protection on the counterparty. This in effect allows one to trade away counterparty credit risk.
(iii) Computation of risk weighted assets and capital requirements.
(iv) Obtaining insight about prices of complex transactions in potential future scenarios. For example, while counterparty risk is concerned with measuring how *high* the value of a transaction can go (and therefore how much a counterparty would owe), there are similarities between this and computing Value at Risk, or stress testing, where one would be interested in how much the value of a transaction could drop.

1.4 Modelling Counterparty Credit Exposure

In the previous sections we have introduced the concept of counterparty exposure and have provided some simple examples. We focus now on a more formal approach which will give a flavour of the mathematical tools we will need in the next chapters.

1.4.1 Definition

Given a portfolio of positions traded with a counterparty, the main quantity we need to model to compute the counterparty credit exposure at time t, is the distribution V_t of the portfolio prices, computed at time $t > 0$ and seen from today. We will see in the next chapter how V_t can be described in its full generality. For the moment we consider the case of products without callability features and where cashflow payments are performed at discrete time points (T_i), $i = 1, \ldots, n$, with T_n being the maturity of the trade. Define X_t to be the (generally stochastic) payment made by the portfolio at any time t ($X_t = 0$ if t is not a member of (T_i), $i = 1, \ldots, n$). Then at any time $t \geq 0$, V_t can be expressed as:

$$V_t = N_t \sum_{T_i > t}^{T_n} \mathbb{E}\left(\frac{X_{T_i}}{N_{T_i}} \,\bigg|\, \mathscr{F}_t\right), \tag{1.1}$$

where N_t indicates the numeraire, $\mathbb{E}$ is the expectation in the numeraire measure and $\mathscr{F}_t$ the usual filtration. More details of the concept of numeraire, pricing measure, and filtration can be found in the literature (see for example Baxter & Rennie [10] for an intuitive description, or Rogers & Williams [93, 94] and Shreve [98] for a more formal approach) and will also be given later in this book (see Appendix B). For our purposes here it is enough to think of the numeraire as being the cash account, used to discount cashflows, and the filtration as the information available at time t.

At time $t = 0$ the distribution degenerates into the current price of the portfolio. We are interested in the distribution of V_t under either the real or the pricing (called also risk-neutral) measure. In general the price distribution V_t will change with time due to changes in market conditions, portfolio composition (for example due to payment of cashflows), and time value. If the portfolio is collateralised, it can be extended to take into account additional positions representing the collateral value. The computation of the price distribution V_t depends also on specific contractual features with the counterparty, e.g. netting agreements between short and long positions in the portfolio, or break clauses held by the counterparties.

The industry practice to compute exposure is to use a simple Monte Carlo framework implemented in three steps: (i) scenario generation, (ii) pricing, and (iii) aggregation.

The first step involves generating scenarios of the underlying risk factors at future points in time. Simple products can then be priced on each scenario and each time step, therefore generating empirical price distributions. From the price distribution at each time it is then possible to extract convenient statistical quantities. Exposure of portfolios can be computed by consistently pricing different products on the same underlying scenarios and aggregating the results taking into account possible netting and collateral agreement with the counterparty.

If taken literally, this approach works only for relatively simple products which can be priced analytically, or which can be approximated in analytical form, and which do not need complex calibrations depending on market scenarios. More exotic products requiring relatively complex pricing, cannot be treated in this way. As already mentioned, even a cancellable swap, which is a relatively simple product, cannot be computed easily in this framework.

In the next chapters we will show how (1.1) can be generalised and which algorithms can be implemented to compute exposure for more exotic products. We will also challenge the simple Monte Carlo approach we have just described, and see how more sophisticated modelling frameworks can provide answers to some of the common problems faced when building a counterparty exposure system.

1.4.2 Risk Measures

For practical reasons it can be useful to characterise the distribution V_t with some statistical quantities which can then be used for various risk controlling or risk management purposes. The Potential Future Exposure (PFE), computed at time t is defined as

$$\text{PFE}_{\alpha,t} = q_{\alpha,t} = \inf\{x : \mathbb{P}(V_t \leq x) \geq \alpha\}, \tag{1.2}$$

where α is the given confidence level, and $\mathbb{P}$ indicates the probability distribution of V_t. Note that this is a function of time t and is the price of the obligation *in the future* given a set of scenarios. This pricing is called sometimes *Mark-to-Future*. The graph of $\text{PFE}_{\alpha,t}$ as a function of t is known as the exposure profile of the trade.

Similarly the Expected Positive Exposure (EPE) will be computed as[9]

$$\text{EPE}_t = \mathbb{E}\left[V_t^+\right], \tag{1.3}$$

where the expectation can be taken under the real or pricing measure depending on the usage of EPE.

An alternative measure to the quantile is the Expected Shortfall, called also Expected Tail Loss, defined as

$$\text{ES}_{\alpha,t} = \mathbb{E}\left[V_t \mid V_t > \text{PFE}_{\alpha,t}\right]. \tag{1.4}$$

Expected shortfall is used especially when it is convenient to have a measure which takes into account events of significant magnitude, which, however, can occur with only very small probability. As we have shown above, typical examples are credit derivatives, where the default of the reference entity protected by the derivative is a low probability event, which, however has significant impact.

1.4.3 Netting and Aggregation

In general, the credit exposure to a particular counterparty arises not from a single transaction but several ones. For any particular market scenario, some of these transactions will have positive, and others negative value. Consider, for example, a long and a short position on an option on highly correlated stocks, a portfolio of payers and receivers swaps[10] in different currencies, or, as a more sophisticated example,

---

[9]We will see later in Chap. 12 and Chap. 14 that other definitions of EPE are more appropriate to compute CVA.

[10]A payer (resp. receiver) swap, is a swap that pays (resp. receives) a fixed rate and receives (resp. pays) a floating Libor rate.

a long position on ABX[11] and a short position on a tranche of pool of MBS. One would expect that, at a given time as one position increases in value, the value of the other position decreases. Since both transactions are facing the same counterparty, it is natural to think about the possibility of *netting* these positive and negative values together, in order to reduce the overall exposure.

The possibility of treating risk in this way will depend on the legal agreement in place. Netting agreements can have different flavours. For example for a given counterparty it could be possible to net together interest-rate swaps, but not swaps with e.g. equity transactions.

From the quantitative and computational perspective netting and no-netting agreements will determine how aggregation is performed within a pool of transactions. The main challenge is the requirement of being scenario consistent across trades. This means that the price distributions of all transactions have to be computed together in order to choose the correct risk measure of the whole portfolio together with the correct netting agreements. This can pose significant constraints on the software architecture as well as on the computational capacity.

Once counterparty exposure is computed at portfolio level, one can be interested in assigning a portion of the exposure to each single transaction. It is interesting to note that this is not equivalent to computing exposure for each single transaction separately. This process of redistributing exposure is often called *exposure allocation* or *disaggregation* and can be performed in different ways leading to different results.

We will analyse quantitative aspects of both aggregation and allocation in Part IV where we discuss hedging and managing counterparty risk.

1.4.4 Close-Out Risk

Close-out risk refers to the possibility of loss during the time period between when a counterparty is deemed to be in default and when the transaction with that counterparty has been wound down or replaced in the market. The length of this time period, referred to as the *close-out period*, is typically assumed to be ten business days. In practice it may be shorter for liquid transactions or longer for specialised and bespoke transactions.

To mitigate close-out risk, a collateral agreement is often included in the transaction. Under such an agreement, the counterparty would have a commitment to post assets (be they in form of cash or other highly-rated assets) whenever the exposure from the transaction is observed to increase. There are several components that may be specified in a collateral agreement, such as (i) an initial upfront collateral amount called the *initial margin*, (ii) the threshold exposure above which extra collateral

---

[11]The ABX Index is a series of credit-default swaps based on 20 MBSs that relate to subprime mortgages.

would need to be posted, (iii) the minimum amount of collateral that may be posted on each collateral exchange date, and (iv) the frequency of the margin calls.

The collateral agreement is a legal agreement also referred to as the CSA (Credit Support Annex). Typically the terms of this agreement will depend on the jurisdiction where it applies. In Part IV we will analyse some quantitative and modelling aspects of close-out risk, without addressing all the intricacies of the legal aspects.

1.4.5 Right-Way/Wrong-Way Exposure

In all the examples we have analysed previously, we did not consider the quality of the counterparty, assuming in effect that counterparty exposure is equivalent to the future replacement value of the trade at time of counterparty default. In general, however, the level of exposure caused by the trade and the quality of the counterparty are not independent of each other. Information about one would force us to re-evaluate information we have about the other. We refer to such dependence as right-way or wrong-way exposure. The question is how to factor this effect into a credit quantification computation. Typical examples where such considerations are called for are when call or put options are written on the counterparty's own stock. These are limiting cases with practically no need of accurate modelling. The problem becomes more interesting when the product is complex and the correlation between counterparty quality and level of exposure cannot be clearly determined.

Consider for example an energy producer which swaps energy futures for a stream of coupons. In general, increases in energy prices could be beneficial to the company, therefore reducing its probability of default. One way of taking into account the Right Way risk is to measure correlation between energy prices and company credit spread.

Another interesting example[12] of wrong way risk are negative basis swaps performed with monoline insurance companies. Typically in this case the insurance company receives a premium and pays default protection on missing payments from a pool of mortgages. These swaps are called negative basis as, in normal market conditions, the price for the protection is lower than the value implied by the spread paid by the mortgages.

1.4.6 Credit Valuation Adjustment: CVA

Once counterparty exposure has been computed it is necessary to find ways of mitigating it. The simplest way is to compare the portfolio PFE with pre-defined limits and constrain the amount of transacted notional or, as we have seen previously, enter

---

[12] ...especially in light of the 2007/2008 credit events....

into a collateral agreement. A possible alternative consists in buying credit protection on the counterparty. Its price corresponds to the value of the protection leg of a CDS that pays the exposure amount in case of default of the counterparty. This value is called in the industry credit valuation adjustment, CVA.

Intuitively we can see this as follows. Within a pricing framework the value of credit exposure can be seen as the expected value of the positive part of the price distribution weighted by the default probability. Assuming that prices are independent from defaults, we can separate expectations, obtaining that CVA is the value of a CDS with the notional being the EPE profile of the underlying transaction. Suppose for simplicity that the EPE profile is a piece-wise constant function over a time interval $(T_i - T_{i-1})$.

$$\mathrm{CVA} = \sum_i \mathrm{EPE}_i (T_i - T_{i-1}) D_{0,T_i} s_i, \tag{1.5}$$

where s_i is the spread corresponding to the time interval $T_i - T_{i-1}$ and D_{0,T_i} the discount bond maturing at time T_i. We can see that the CVA corresponds to a portfolio of forward starting CDSs (or equivalently long and short CDS positions) with piecewise constant notional. The availability of CDSs of different maturities will dictate how the EPE profile is discretized.

The CVA depends on the level of exposure as well on the credit spread of the counterparty. As counterparty exposure and spread change with time, the amount of credit protection needs to be adjusted accordingly. The process of balancing of exposure with CDSs and other instruments sensitive to market parameters corresponds to dynamically hedging counterparty credit exposure. More details on how to compute and hedge CVA are given in Chap. 14.

1.4.7 A Simple Credit Quantification Example

We will discuss in detail in Chap. 9 the computation of credit exposure for equity products. We consider here a very simple example where the form of the exposure profile and the maximum values of the PFE can already be deduced from an approximation.

Suppose company A has bought from counterparty B a call option of strike K on a stock S. Our goal is to compute the credit exposure and close-out risk company A is facing. As mentioned in the previous section we need to calculate the price distribution V_t. In the case of the call option, in a simplified context where rates are deterministic, (1.1) becomes

$$V_t = N_t \mathbb{E}\left[\frac{(S_T - K)^+}{N_T} \middle| \mathscr{F}_t \right] = e^{-r(T-t)} \mathbb{E}\left[(S_T - K)^+ \mid \mathscr{F}_t \right], \tag{1.6}$$

where S is the stock price, K the strike, and r the interest rate assumed to be constant. The notional (number of options) has been assumed equal to one. As mentioned previously, $\mathscr{F}_t$ is the usual filtration, N_t the numeraire, and the expectation is taken in the measure $\mathbb{N}$.

To solve this equation we need for the stock price S a model, with which simulate the stock value till maturity T. A simple model, which is often used in credit, is the geometric Brownian motion with constant volatility σ, interest rate r and dividend yield d.

$$\frac{dS}{S} = (r - d)dt + \sigma dW_t, \tag{1.7}$$

where W_t is a standard Brownian motion. As it is well known this stochastic differential equation can be solved analytically.

Thus, to compute exposure we need to simulate the stock with (1.7), and then, using the Black and Scholes formula [15] we can price the option at each time step and in each scenario.

As the exposure of an equity option is generally monotonic in the underlying and is growing with time, and a vanilla stock option depends only on the current stock value (the product is not path dependent), the max PFE will be in general at maturity T.[13] We can compute it at let's say 97.5% confidence level as (see also Appendix A)

$$\text{PFE}_T = S_0 \left(e^{(r-d-\frac{\sigma^2}{2})T+1.965\sigma\sqrt{T}} \right) - K, \tag{1.8}$$

assuming it is a positive quantity. The expected exposure can be computed in this simple case as

$$\text{EPE}_t = V_0 e^{rt}, \tag{1.9}$$

where V_0 is the option premium. We can see this as following. Given that the value of a call option is always positive, we can write (in our simplified set-up),

$$\text{EPE}_t = \mathbb{E}[V_t^+] = \mathbb{E}[V_t] = \mathbb{E}[\mathbb{E}[e^{-r(T-t)}(S_T - K)^+ | \mathcal{F}_t]] = V_0 e^{rt}. \tag{1.10}$$

As for approximating the close-out exposure for a short close-out period, one can use a first-order Taylor approximation.

$$\text{CloseOut}_t \approx V_0 + \Delta(S_t - S_0), \tag{1.11}$$

where Δ is the first order derivative of the call option price with respect to the stock and S_t is the value of the stock during the close-out period. This is the close-out risk for the initial period, i.e. for the time between time-zero and t. We will see later in this book that the computation of close-out risk presents subtleties which go far beyond this simple computation.

---

[13]The exact shape of the PFE curve will depend on the interest rate, dividend curve, and option characteristics.

1.4.8 Computing Credit Exposure by Simulation

Within a Monte Carlo framework, to compute exposure we could simulate the stock price from today to maturity using (1.7), and then price the option on each path and each time step using Black and Scholes, again with constant rate, dividend, and volatility. As we will see in the next chapters, it is more convenient to simulate martingale processes, for which only the volatility structure is relevant, while the drift (and thus the dividends) does not need to be specified (for a definition of martingale see Appendix B and for more details see for example the books by Baxter & Rennie [10], Rogers & Williams [93, 94] and Shreve [98]). In practice a convenient quantity we can simulate are forward prices. By considering our example in these terms, we can write (1.6) as

$$V_t = N_t \mathbb{E}\left[\frac{(F_{T,T} - K)^+}{N_T} \middle| \mathscr{F}_t \right],$$ (1.12)

where $F_{t,T}$ is the t-value of the T-forward. The link with the notation in the previous section is,

$$F_{t,T} = S_t e^{(r-d)(T-t)}.$$ (1.13)

As before, assume for simplicity that the numeraire N is independent from the stock price, and impose also the simple specification

$$dF_{t,T} = F_{0,T} \sigma \, dW_t,$$ (1.14)

with the volatility being a constant $\sigma > 0$ and with W being a standard $\mathbb{N}$-Brownian Motion. This is the quantity we simulate in t. The paths generated by integrating this SDE are our scenarios which we show in Fig. 1.5 in a stylised representation.

Note that (1.14) can be integrated analytically at each time step, thus avoiding discretisation errors,

$$F_{t,T} = F_{0,T} \exp\left(-\frac{\sigma^2}{2} t + \sigma W_t \right).$$ (1.15)

We can then price at each scenario and each time step the stock option using again Black and Scholes, expressed in terms of the forward price $F_{t,T}$.

PFE can be computed analytically at maturity, where the option price is given by the stock price minus the strike.

$$\text{PFE}_T = \left[F_{0,T} \exp\left\{ \sigma \sqrt{T} \tilde{q}_{\alpha,T} - \frac{1}{2}\sigma^2 T \right\} - K \right]^+,$$ (1.16)

where $\tilde{q}_\alpha = \Phi^{-1}(\alpha)$ is the α-quantile of the standard normal distribution. This is the equivalent of (1.8) generalised for any quantile. We have also floored at zero the exposure, as in some cases one is interested only at the amount the counterparty should pay. Negative exposure represents the amount we owe to the counterparty.

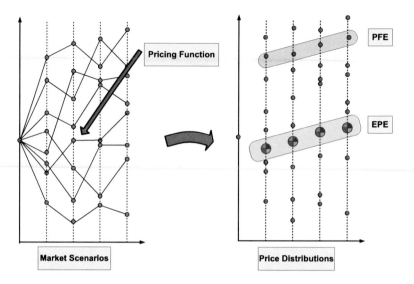

Fig. 1.5 Computing exposure by Monte Carlo simulation. The paths on the left panel represent stock prices. At each scenario and each time step, the price of the option is computed using the analytical Black and Scholes formula. Resulting prices are represented on the right panel. From the price distributions generated in this way at each time step, various statistical quantities (e.g. PFE and EPE) can be extracted. The bigger circles indicate a mean

1.4.9 Implementation Challenges

The Monte Carlo framework we have shown in the previous section, seems to give a good implementation recipe. For a given portfolio of transactions we could (i) identify the underlying risk factors and simulate forward (or spot) prices, taking into account correlations if required, (ii) use functions already implemented to price each product, and then (iii) derive statistical quantities. As we have mentioned already, this could be the approach followed by a financial institution to assess the counterparty credit risk of its OTC derivatives portfolios.

In the implementation phase, however, there can be issues which need to be addressed.

(i) The generation of correlated scenarios is not trivial, as there can be thousands of different risk factors driving the dynamics of products in the portfolio. Consider for example an equity portfolio, where each underlying stock needs, at least in principle, a specific simulation.

(ii) The scenarios have to be consistent across systems to build a counterparty view. This is a requirement which is much more stringent than what is generally specified in the design of a Front Office system used for pricing or a Risk system used to monitor the Profit and Loss (P&L) of a bank. Basically what we need here is the same underlying models, or the same family of models, for all types of products. In fact, even if the correlation between asset classes can be in some cases ignored (e.g. equity could be considered not correlated with interest rate), still all these models need to be expressed using the same numeraire (the discount factors in equity have to be consistent with the discount factors used in FX or rates). This consistency can be difficult to achieve, as often large financial companies have different systems to book and value, for example, interest-rate, equity, or FX products.

(iii) Pricing functions developed in various libraries are not necessarily designed to be integrated in a counterparty exposure framework. This has implications from both a software and architecture, as well as from a methodological point of view. Consider for example path dependent products. Counterparty exposure depends on the whole scenario history, which could be in different formats across different pricing systems.

(iv) Not all products can be computed in analytical form. Most exotics are priced on grids using PDEs or using Monte Carlo approaches. In these cases the exposure computation would require a Monte Carlo simulation for scenarios and a Monte Carlo simulation, or a PDE computation, for each scenario and time step to price the instrument. This becomes quickly unfeasible from a computational point of view. In addition, depending on the model used for pricing, calibration could also become problematic, as it has to be performed at each scenario.

In practice credit systems based on the classical Monte Carlo scheme approximate products using a simplified representation. While these approximations could have their justification in a risk environment, they are difficult to use when counterparty risk has to be priced and hedged.

1.4.10 An Alternative Approach: The AMC Algorithm

The points highlighted in the previous section clearly show that the classical Monte Carlo scheme has intrinsic limitations and that we need an alternative approach. As we will see at length in the rest of the book, there are possibilities to circumvent in a systematic way some of the problems related to valuation and architecture.

The basic idea is to approach the counterparty exposure problem as a pricing problem, and thus to use *pricing* algorithms, which generate not just the value of a trade at inception, but rather a price distribution at predetermined time steps. One possibility is to use the so called American Monte Carlo algorithm, which we will refer to as, simply, the AMC algorithm. The main feature of this algorithm is that, instead of building a price moving forward in time, it starts from maturity, where the value of the transaction is known, and goes backwards, till the inception.

In general AMC is used for pricing products with callability, i.e. products whose values depend on a strategy which can be determined by only knowing future states of the world. From a counterparty exposure perspective, the benefit of this approach is that, not only a price at time-zero is provided, but also the price distribution at each time step. In addition, the algorithm is generic, in the sense that using simply a payoff description, we can obtain the information needed to compute counterparty

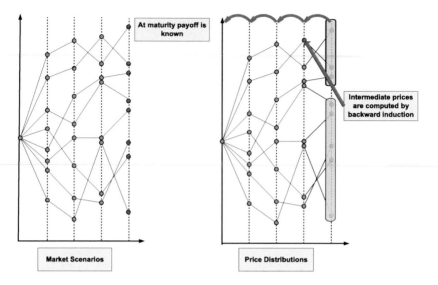

Fig. 1.6 A simplified graphical representation of the AMC algorithm. In the left hand panel we show the scenarios generated according to some underlying model. At maturity the payoff of the trade is known. To estimate the value at intermediate scenarios we need to proceed with a backward induction step

exposure. This suggests the possibility of having a generic trade representation and thus the possibility of having a modular software architecture that incorporates trade descriptions without explicit knowledge of each type of product. The challenge is that we need to develop an underlying model capable of pricing a *hybrid product*, consisting potentially of a large portfolio of transactions. This hybrid model will need to take into account all stochastic drivers of the portfolio in a consistent, arbitrage free way.

It is natural to ask what is the performance of the AMC algorithm for vanilla products. We will see that by a careful implementation the prices computed via AMC are very close to those computed using for example closed form formulae.

1.5 Which Architecture?

Building a system that computes credit valuation adjustments and counterparty exposure for the book of a large financial firm is a very challenging task, not only from the modelling and algorithmic perspective, but also from the technical and IT point of view. One of the problems is that often in large institutions such as Investment Banks, products are not booked on one system. They are in general recorded on several systems, which do not necessarily communicate between each other. To overcome this situation we suggest developing not only a common modelling platform, but also a programming language, which allows the representation of different types of products. As we will see later in this book, we have called our language

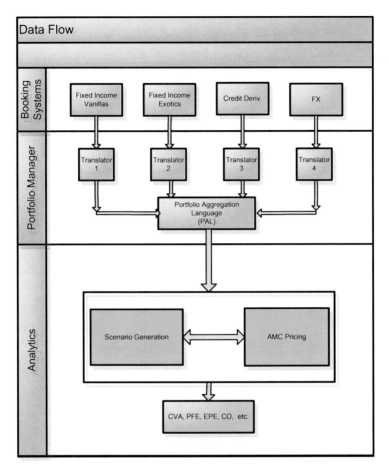

Fig. 1.7 High level architecture description

PAL, Portfolio Aggregation Language, to highlight the fact that we need to aggregate trades at counterparty (or at netting pool) level.

Once we have this common booking language, we can translate bookings made in other systems into PAL, bridging the difference between these systems. In the figure below we show how the system architecture could be implemented.

1.6 What Next?

We have introduced all basic concepts needed to understand counterparty credit exposure. We have now to analyse in detail the steps necessary to build a system designed to compute and hedge counterparty risk for large portfolios of exotic transactions. This is what we do in the rest of this book. We start from a generic modelling and simulation framework based on American Monte Carlo techniques, and then we present a software architecture, which, with its modular design, allows the computation of credit exposure in a portfolio-aggregated and scenario-consistent way.

Chapter 2
Modelling Framework

Our goal is to define a general framework which can be used to compute counterparty credit exposure for all types of transactions. As highlighted in the Introduction, computing counterparty exposure consists of computing distributions of prices at future times. For simple products this can be achieved by scenario simulation, followed by pricing on each scenario, at each time step. However, in the case where no analytical form is known for the price of the product, this approach is not practical and a different approach is required.

The framework we define is intended to cater for the need to estimate price distributions in a consistent way, not just for a single product, but for complex portfolios of products admitting no closed-form pricing. Price distributions will be obtained through simulating underlying processes specified in a generic way, and this chapter is dedicated to defining the problem at hand and describing the various processes to be simulated. There are several ways to model the stochastic processes driving the price of derivatives. We start from a generic representation and show how to adapt it to cope with practical constraints. By making certain independence assumptions, we then analyse how to modify the model so that new asset classes can be introduced in a modular way that maintains scenario consistency across products *but does not require the re-evaluation of models already in place*. Such assumptions have consequences for pricing, and these are assessed in Chap. 6.

After having prescribed the theoretical framework, the next chapters will be focused on implementation. In Chap. 3 we will specify simulation models. For vanilla trades, which can be priced analytically, this will be the main step to compute credit exposure, as there is a direct mapping between scenarios of the underlying risk factors and price distribution. In Chap. 4 we will analyse valuation techniques that can be applied when this mapping is not obvious, as, for example, with products having callability features and for which no closed-form valuation is available.

2.1 Counterparty Credit Exposure Definition

Consider a financial product, which we denote generically by P, and denote by $X \equiv (X_t)_{0 \leq t \leq T_X}$ the cashflows which the holder of the product is entitled to. In

G. Cesari et al., *Modelling, Pricing, and Hedging Counterparty Credit Exposure,*
Springer Finance, DOI 10.1007/978-3-642-04454-0_2,
© Springer-Verlag Berlin Heidelberg 2009

general, P might entitle the holder to not just the cashflow X, but also the option to replace, at predetermined points in time, the cashflow X with an alternative product, Q say. Associated with the product Q are cashflows Y. We also generally assume in the following that *the option allows exercise only once.*

Now let τ_E^* denote the optimal (stochastic) time at which the option should be exercised to replace the cashflows X with the product Q. The basic problem at hand is to compute at all times t the value of the product P *whose payoff may depend on decisions made after time t.* The fair value for product P should be that attainable by employing the optimal among all possible exercise strategies τ_E. Solving the pricing problem entails solving for the best possible exercise strategy.

We now formalise what we have said above with a definition, and introduce the required notation that will serve us in Chap. 4 when we describe valuation techniques in detail. First, denote by $\mathscr{T}$ the set

$$\mathscr{T} := \{\tau_1, \tau_2, \ldots, \tau_{n_E}\} \cup \{\infty\} \tag{2.1}$$

of points where the holder of the option is allowed to exercise from P into Q. Even if the optionality is of American type and exercise is allowed at any t within a time interval, any attempt at numerical solution would need to approximate the problem with one where exercise is allowed at a discrete set of points. The point at infinity is included in $\mathscr{T}$, as this corresponds to the possibility of the holder never deeming it optimal to exercise into Q.

Write $V^Q \equiv (V_t^Q)_{t \geq 0}$ for the value process[1] of the product Q, and let N denote our chosen numeraire process. At any time t before exercise, the holder of the product P chooses the optimal exercise time $\tau_E^* \in \mathscr{T}$ so as to attain,

$$V_t^P = N_t \sup_{\tau_E \in \mathscr{T}_t} \left\{ \mathbb{E}\left[\int_t^{\tau_E \wedge T_X} \frac{X_u}{N_u} du \,\bigg|\, \mathscr{F}_t \right] + \mathbb{E}\left[\frac{V_{\tau_E}^Q}{N_{\tau_E}} \,\bigg|\, \mathscr{F}_t \right] \right\} \quad (t < \tau_E^*), \tag{2.2}$$

where $\mathscr{F}_t$ is the usual filtration at time t, $\mathscr{T}_t = \{\tau \in \mathscr{T} \mid \tau \geq t\}$ is a subset of $\mathscr{T}$, and where the expectation is taken in the pricing measure relating to the numeraire N. For vanilla products without callability features, we have $\mathscr{T} = \{\infty\}$, so that the question of optimising the time of exercise does not arise.

Here on the left hand side of (2.2), V_t^P is the time-t value of P *conditional on exercise not having happened prior to t.* This is the *pre-exercise value* of product P. On the right-hand side, the two terms,

$$\Pi_t^{no} := N_t \mathbb{E}\left[\int_t^{\tau_E \wedge T_X} \frac{X_u}{N_u} du \,\bigg|\, \mathscr{F}_t \right]$$

$$V_t^Q = N_t \mathbb{E}\left[\frac{V_{\tau_E}^Q}{N_{\tau_E}} \,\bigg|\, \mathscr{F}_t \right], \tag{2.3}$$

---

[1] Throughout we work with processes defined on a filtered probability space $(\Omega, \mathscr{F}, (\mathscr{F}_t)_{t \geq 0})$. More details are given in Appendix B.

represent, respectively, the time-t value of the non-exercise portfolio containing the X cashflows,[2] and the time-t value of entering into the exercise portfolio containing the Y cashflows at time τ_E.

Upon exercise at τ_E^*, the product will have value $V_{\tau_E^*}^Q$. The credit exposure at times $t > \tau_E^*$ will depend on the way in which the option inherent in P may be exercised. We will discuss in Chap. 4 several types of exercise that are possible. Here it suffices to point out that exercising into Q might give rise to credit exposure for the lifetime of the product Q, which may exceed the maturity T_X of the pre-exercise flows X.

Putting everything together, we may write the price of product P as

$$V_t = \begin{cases} V_t^P, & t < \tau_E^* \\ V_t^Q, & t \geq \tau_E^*. \end{cases} \tag{2.4}$$

In a complete market, there will be unique prices for any derivative that can be replicated. In particular, in a simulation framework, the price calculated will not depend on the numeraire (and therefore the numeraire measure) chosen for the simulation. However, the *distribution* of prices that drives the credit exposures *will* depend on the simulated processes, which in turn depend on the measure chosen for the simulation. Thus, for instance, there could be good reasons for generating scenarios in the physical measure even if pricing is accomplished by taking expectations in the numeraire measure corresponding to numeraire N. We will analyse this point in more detail in Part IV, where we discuss Hedging and Managing Counterparty Risk.

From (2.2), it is apparent that the elements required to simulate values of the distribution V_t are

(i) Simulations of the numeraire process N.
(ii) The ability to simulate the cashflows X provided by P, and similarly the cashflows offered by Q. Since all these cashflows are usually stochastic, we need to specify a simulation model tailored to the particular products P and Q. This is what we deal with in Chap. 3.
(iii) Estimation of the price of the no-exercise portfolio Π^{no} and that of the product Q for each point in time at which we are interested in having a simulated value available. Similarly, we need to account for the sup operator numerically within the simulation. We will consider this aspect in Chap. 4.

The rest of this chapter is dedicated to describing our general approach to modelling the dynamics of the numeraire and the cashflows of the generic products P and Q mentioned in (i) and (ii) above.

---

[2]Note that X in general can be decomposed as $X_t = X_t^c + \sum \delta_{t_j} x_j$, where X^c is a continuous process, x_j is a cashflow at time t_j and δ_t is the Dirac delta function.

2.2 Process Dynamics

Most approaches to simulating the financial quantities required for credit exposure estimation choose to model stochastic variables that are directly observable, e.g. Libor rates, swap rates, stock prices, foreign exchange rates. While specific payoffs can usually suggest which observable quantity it is more sensible to simulate, there is the inherent drawback that a new payoff may require simulation of new quantities, which the framework in place had not been designed to handle.

We take an alternative approach, for which

(i) Extraction of all stochastic quantities of interest (for all possible payoffs) is done from the *same* basic set of martingale[3] processes. This set does not need to be augmented in order to tackle a new payoff type.
(ii) Time-zero term structures are reproduced *exactly by construction*. The specification of a new model is therefore reduced to specifying the volatility structure. Such specification must allow for fast and accurate calibration to vanilla option market prices.

The basic idea is as follows. Suppose we have a contingent claim X_T paying off at time T. The risk-neutral price at $t < T$ of X_T is given by[4]

$$p_0 = \mathbb{E}^{\mathbb{N}} \left[(N_T/N_0)^{-1} X_T \right], \tag{2.5}$$

the expectation being in a measure that makes martingales out of (non-dividend-bearing) tradable assets when expressed in units of the numeraire N. We can equivalently use the T-bond as numeraire, so that

$$p_0 = \mathbb{E}^{\mathbb{N}^T} \left[(D_{T,T}/D_{0,T})^{-1} X_T \right]$$
$$= D_{0,T} \mathbb{E}^{\mathbb{N}^T} [X_T]$$
$$= D_{0,T} \mathbb{E}^{\mathbb{N}} [M_{T,T} X_T], \tag{2.6}$$

where the expectation is now in a measure where prices of (non-dividend bearing) assets expressed in units of the T-bond are martingales. The usefulness of this second representation is that by modelling the change-of-measure process[5]

$$M_{t,T} \equiv \left. \frac{d\mathbb{N}^T}{d\mathbb{N}} \right|_{\mathscr{F}_t} = \frac{D_{t,T}/D_{0,T}}{N_t/N_0}, \tag{2.7}$$

---

[3]A martingale is a stochastic process whose expected future value, conditional on present information, is its current value. In mathematical terms this can be expressed as $\mathbb{E}[M_t \mid \mathscr{F}_s] = M_s$, for all $s \le t$ (see Appendix B for definitions and, for example, Williams [106], or Shreve [98] for a detailed treatment of martingale theory).

[4]For clarity we explicitly indicate the measure in which expectation is taken; in the rest of the book we will ignore the superscript $\mathbb{N}$ when expectation is taken in the numeraire measure $\mathbb{N}$.

[5]See Appendix B for more details.

the time-zero price of X_T is a multiple of time-zero observed bond prices. That is to say, the yield curve at time zero is replicated by construction. Comparing (2.5) and (2.6), we see moreover specifying the martingale $M_{.,T}$ forces the numeraire to take the form

$$N_T^{-1} = D_{0,T} M_{T,T}.$$ (2.8)

At a general level, our approach is similar to that followed in Constantinides [28], Flesaker [45], and Rogers [90]. These papers model directly (in some fixed reference measure $\mathbb{P}^*$) the state-price density process ζ, which has the property that for any non-dividend-bearing asset with price process S,

$$\zeta_t S_t \text{ is a } \mathbb{P}^*\text{-martingale.}$$ (2.9)

If the numeraire process is increasing, then ζ is a positive supermartingale. It is this property that Flesaker in [45] and Rogers in [90] exploit to come up with a surprising variety of interest-rate models, essentially writing

$$\zeta_t = e^{-\alpha t} f(X_t)/f(X_0)$$ (2.10)

for a given Markov process X and inspired choices of the function f. We choose to work in the reference measure $\mathbb{P}^* = \mathbb{N}$ and to model the numeraire directly.

The recipe for modelling assets other than the chosen reference currency (foreign currencies, equity and inflation) is similar and will be elaborated on further below. Conceptually, there will be two new sources of randomness introduced with each new asset class. The first of these is the output derived from owning one unit of the asset (bond yield volatility in the case of a currency, and stock dividend volatility in the case of equity), and is modelled by introducing for each asset i a martingale process $M_{.,T}^{(i)}$, analogous to (2.7) above. The second source of randomness is the fluctuation of the asset price itself (that is to say, the exchange rate process in the case of a currency foreign to the reference one, and the stock price process in the case of equity) when expressed in some existing unit of value; this will be modelled through a second process, which we typically denote $Y^{(i)}$ for the i'th asset. Credit derivatives present different challenges; nevertheless, the idea there will still be to find the relevant martingale and to directly specify dynamics for it in our chosen reference measure.

2.3 Interest Rate: Single Currency

We start by considering the modelling of the reference currency itself. All the ingredients for this have been given by way of motivation in Sect. 2.2, namely the change-of-measure martingales (2.7)

$$M_{t,T} = \frac{D_{t,T}/D_{0,T}}{N_t/N_0} \quad (t \le T)$$ (2.11)

which in fact can also be interpreted as the price process of the T-bond, expressed in units of the numeraire N (for $t > T$ we assume $M_{t,T} = M_{T,T}$).

We will take $M_{t,T}$ as the fundamental quantity to be simulated and from which the other interest rates-related stochastic quantities we need can be extracted.

Table 2.1 Single currency interest-rate product notation

$D_{t,T}$	T-bond in the local (domestic) currency
N_t	Numeraire in the local currency
$M_{t,T}$	Martingale used to simulate the T-bond in the local measure
$\mathbb{N}$	Measure in which prices of non-dividend-bearing assets expressed in units of N are martingales
$\mathbb{N}^T$	Measure in which prices of non-dividend-bearing assets expressed in units of the T-bond are martingales

While, for each t, N_t depends only on the "diagonal" values $M_{t,t}$ of the martingale $M_{\cdot,t}$, it is clear that simulated values of $M_{t,T}$ will be required for all $t \in [0, T]$, and for each $T \geq 0$. Indeed, from (2.11) it follows that

$$\frac{M_{t,T}}{M_{t,t}} = D_{t,T}\frac{D_{0,t}}{D_{0,T}}, \tag{2.12}$$

which means that the time-t price of a T-bond is

$$D_{t,T} = \frac{D_{0,T}}{D_{0,t}}\frac{M_{t,T}}{M_{t,t}}. \tag{2.13}$$

Other quantities of interest, such as lending rates, can similarly be written in terms of the martingales $M_{t,T}$.

All that is left now is to choose an SDE for the strictly positive martingale M. Consider, for example, for each $T \geq 0$ the SDE

$$dM_{t,T} = M_{t,T}\sigma_{t,T}dW_{t,T}, \tag{2.14}$$

where $W_{\cdot,T}$ is a Brownian Motion process and $\sigma_{t,T}$ is a volatility term. For deterministic σ, (2.14) can be written in integral form as

$$M_{t,T} = M_{0,T}\exp\left(\int_0^t \sigma_{u,T}dW_{u,T} - \frac{1}{2}\int_0^t \sigma_{u,T}^2 du\right), \tag{2.15}$$

where

$$t\Sigma_{t,T}^2 := \int_0^t \sigma_{u,T}^2 du \tag{2.16}$$

is the total variance of $\ln M_{t,T}$.

The framework described by (2.15) gives the simulation recipe: first choose a volatility structure σ, then simulate the martingale $M_{t,T}$ and finally derive the stochastic quantities which are needed to estimate price distributions of the relevant product. For example, the numeraire can be written in integral form (for deterministic volatilities) as

$$N_t^{-1} = D_{0,t} M_{t,t}$$

$$= D_{0,t} M_{0,t} \exp\left\{ \int_0^t \sigma_{u,t} dW_{u,t} - \frac{1}{2} \Sigma_{t,t}^2 \right\}. \tag{2.17}$$

Similarly, the Libor rate of interest $L_{t,[T_1,T_2]}$ observed at t for the period $[T_1, T_2]$, is (with $\alpha \equiv T_2 - T_1$)

$$L_{t,[T_1,T_2]} = \alpha^{-1} \left(\frac{D_{t,T_1}}{D_{t,T_2}} - 1 \right)$$

$$= \alpha^{-1} \left(\frac{D_{0,T_1} M_{t,T_1}}{D_{0,T_2} M_{t,T_2}} - 1 \right)$$

$$= \left(L_{0,[T_1,T_2]} + \alpha^{-1} \right) \frac{M_{t,T_1}}{M_{t,T_2}} - \alpha^{-1}, \tag{2.18}$$

or, in other words,

$$\left(L_{t,[T_1,T_2]} + \alpha^{-1} \right) M_{t,T_2} = \left(L_{0,[T_1,T_2]} + \alpha^{-1} \right) M_{t,T_1} \tag{2.19}$$

showing that simulation of the martingale process and the knowledge of the initial yield curve allows simulation of Libor rates.

2.3.1 Simple Specifications

We have chosen our simulation approach to best serve our goal of computing exposure for all types of products across all asset classes in a consistent way. More familiar models from the literature usually take a different approach, and it is instructive to see how these models can be represented within our framework. This is what we do in this section.

The first practical problem we need to tackle is of course that of dimensionality. In general, the cashflows that need to be simulated for a particular product will depend on values of the martingale $M_{t,T}$ for arbitrary values of T. For example, we see from (2.18) that the Libor rate for two different periods will depend on two different martingales. This means that *all martingales for all values of T are needed* if one is to ensure that cashflows of any product can be extracted. In practice, of course, only finitely many stochastic drivers can be simulated, so one needs to find ways of projecting the richness provided by the infinite family of martingales $M_{.,T}$ onto a finite set of Brownian Motions which can be simulated.

The richness retained by such a dimensional reduction will depend on how many different Brownian Motion processes one chooses. In particular, one can force the martingales $M_{.,T}$ to all depend on the *same* Brownian Motion, as would be the case in any one-factor short rate model. In this case (2.14) becomes,

$$dM_{t,T} = M_{t,T}\sigma_{t,T}dW_t, \quad 0 \le t \le T. \tag{2.20}$$

Note that the $M_{t,T}$ depends on t but also, thanks to $\sigma_{t,T}$, on T. In fact, if σ (and therefore M) were to not depend on T, then bond prices in the model would be deterministic, because the bond price

$$D_{t,T} = \frac{D_{0,T}}{D_{0,t}}\frac{M_{t,T}}{M_{t,t}} = \frac{D_{0,T}}{D_{0,t}}, \tag{2.21}$$

would be determined at time zero. Similarly, for the forward rate one would get

$$f(t,T) = -\frac{\partial \log D_{t,T}}{\partial T} = -\frac{\dot{D}_{0,T}}{D_{0,T}}, \tag{2.22}$$

where $\dot{D}$ represents derivative with respect to maturity of the bond.

From the point of view of implementation, a *separable* specification of the volatility term $\sigma_{t,T}$ in (2.20) turns out to be very useful. In detail, write

$$\sigma_{t,T} = f_t g_T \tag{2.23}$$

for positive deterministic functions f and g. The integral form of $M_{t,T}$ then becomes,

$$M_{t,T} = M_{0,T} \exp\left(\int_0^t \sigma_{u,T} dW_{u,T} - \frac{1}{2} \int_0^t \sigma_{u,T}^2 du \right)$$

$$= M_{0,T} \exp\left(g_T X_t - \frac{1}{2}\Sigma_{t,T}^2 t \right), \tag{2.24}$$

where

$$X_t := \int_0^t f_u dW_u$$

$$\Sigma_{t,T}^2 := t^{-1} \int_0^t f_u^2 g_T^2 du \equiv F_t^2 g_T^2. \tag{2.25}$$

This specification is simple to extend to a multi-factor setting as follows. Starting from an $\mathbb{R}^n$-Brownian Motion $\mathbf{W}$, we write[6]

$$dM_{t,T} = M_{t,T} \mathbf{g}_T \cdot (\mathbf{f}_t \mathbf{R} d\mathbf{W}_t), \tag{2.26}$$

---

[6]Throughout this work we use boldface to indicate vectors and matrices. The dot (scalar) product between two vectors $\mathbf{a}$ and $\mathbf{b}$ is indicated by $\mathbf{a} \cdot \mathbf{b}$, and the row by column product between two matrices (or a matrix and a vector) $\mathbf{X}$ and $\mathbf{Y}$, by $\mathbf{XY}$. The transpose of matrix $\mathbf{X}$ is $\mathbf{X}^\mathsf{T}$. We indicate with $\mathbf{X}^{(i)}$ (or $\mathbf{X}_i$) the i-th column (vector) of matrix $\mathbf{X}$, and with a_j the j-th element (scalar) of vector $\mathbf{a}$. $X_{i,j}$ is the i, j element of matrix $\mathbf{X}$.

where now

$$\mathbf{f}_t = \mathrm{Diag}(f_t^{(1)}, \ldots, f_t^{(n)}) \tag{2.27}$$

is a diagonal matrix with the $f_t^{(i)}$ on the diagonal,

$$\mathbf{g}_T = (g_T^{(1)}, \ldots, g_T^{(n)}) \tag{2.28}$$

is the vector of the $g_T^{(i)}$, and $\mathbf{R}$ is such that $\mathbf{R}^\mathsf{T}\mathbf{R}$ is a positive semi-definite matrix making the i'th component of $\mathbf{RW}$ a standard Brownian Motion, which then gets time-changed by $f_t^{(i)}$. We will see in Chap. 3, for instance, that the Hull-White model is a special case of a parametrization of this type.

The impact of separability is that simulation of *a small number* of Brownian Motions

$$X^{(i)} = \sum_j \int_0^t f_u^{(i)} R_{i,j} dW_u^{(j)} \tag{2.29}$$

together with parametrization of the volatility function $\Sigma_{t,T}^{(i)} = F_t^{(i)} g_T^{(i)}$, allows simulation of $M_{t,T}$ for all T. Without a separable specification for σ, the integral $\int_0^t \sigma_{u,T} dW_u$ would need to be computed on the fly, or stored, for each maturity T. Because of its simplicity, we will employ the separable volatility specification in most of what follows in this book.

2.3.2 HJM Framework

The well-known HJM framework [60] can also be accommodated within our framework, as we will see in the following. In the standard way of specifying an HJM model with $d > 0$ factors, one takes the forward rates $f_{t,T}$ defined via (2.22) and writes down the SDE

$$df_{t,T} = \alpha_{t,T} dt + \mathbf{s}_{t,T} \cdot d\mathbf{W}_t, \tag{2.30}$$

where $\mathbf{W}$ is an $\mathbb{R}^d$-Brownian Motion and $\mathbf{s}_{t,T} \in \mathbb{R}^d$ is a vector specifying the instantaneous volatility at time t of the T-forward rate.

Given that, in our framework, we model directly the bond price $D_{t,T}$, it is natural to compare the bond price dynamics in our framework to those implied by the HJM dynamics (2.30). Indeed, using Itô calculus and (2.30), we can see that in the HJM parametrization the discounted log-bond price satisfies

$$d[\log D_{t,T} N_t^{-1}] = \mathbf{S}_{t,T} \cdot d\mathbf{W}_t - \left(\int_t^T \alpha_{t,u} du \right) dt, \tag{2.31}$$

with

$$\mathbf{S}_{t,T} = -\int_t^T \mathbf{s}_{t,u} du. \tag{2.32}$$

The drift term α in (2.30) cannot be chosen without restriction, as doing so might introduce arbitrage between bonds of different maturities. Arbitrage will be excluded, however, if there exists a process $\gamma \equiv (\gamma_t)_{t \geq 0}$ in $\mathbb{R}^d$ such that

$$\alpha_{t,T} = \mathbf{s}_{t,T} \cdot [\gamma_t - \mathbf{S}_{t,T}], \tag{2.33}$$

as this will guarantee that bond prices discounted by the numeraire are martingales. Notice the key fact that the process γ should make (2.33) true *for all T simultaneously*.

In our framework, the drift condition (2.33) is in fact satisfied automatically. With hindsight this should be expected, since our model laid down at the outset that the discounted bond prices are martingales. To see this, recall that

$$D_{t,T} = \frac{D_{0,T}}{D_{0,t}} \frac{M_{t,T}}{M_{t,t}}. \tag{2.34}$$

Hence, $d[\ln D_{t,T} N_t^{-1}] = d[\ln M_{t,T}]$, and

$$d[\log D_{t,T} N_t^{-1}] = \sigma_{t,T} \cdot d\mathbf{W}_t - \frac{1}{2} \sigma_{t,T} \cdot \sigma_{t,T} dt, \tag{2.35}$$

showing that the volatilities of forward rates within our framework satisfy

$$\mathbf{s}_{t,T} = -\frac{\partial}{\partial T} \sigma_{t,T}. \tag{2.36}$$

2.3.3 Libor Market Models

Libor Market Models, also referred to as BGM models, choose to model Libor rates directly, $L_{t,[T_1,T_2]}$ (for the original description of the model see [17]; for a comprehensive Libor market model analysis see for example Rebonato [89]).

In essence, BGM models require a discretization of the set $\{T_i\}$ of Libor fixing dates. Recall that, at time $t < T_{i-1}$, the Libor rate for time period $[T_{i-1}, T_i]$ is defined by

$$L_{t,[T_{i-1},T_i]} = \alpha_i^{-1} [D_{t,T_{i-1}} - D_{t,T_i}] D_{t,T_i}^{-1}, \tag{2.37}$$

where $\alpha_i = T_i - T_{i-1}$. Since the term in brackets is the difference of two bond prices,

$$L_{t,[T_{i-1},T_i]} D_{t,T_i} \tag{2.38}$$

is a tradable process. Consequently, the process $L_{t,[T_{i-1},T_i]}$ will be a martingale under $\mathbb{N}^{T_i}$, which is the measure specified by taking the T_i-bond as numeraire. This is the cornerstone of the Libor Market Models, which allows the application of the Black formula for pricing caplets by modelling Libor rates as log-normal stochastic processes.

In our framework, the SDE satisfied by the Libor rate is cumbersome because our starting point is martingale modelling of the bond prices. Nevertheless, the SDE can be written down, as we do now. Let us consider, instead the shifted Libor rate

$$\bar{L}_{t,[T_{i-1},T_i]} := L_{t,[T_{i-1},T_i]} + \alpha_i^{-1}. \tag{2.39}$$

From (2.19), we can express M_{t,T_i} in terms of the Libor rates as

$$M_{t,T_i} = \prod_{j=1}^{i} \frac{\bar{L}_0^j}{\bar{L}_t^j} M_{t,T_0} = \prod_{j=1}^{i} \frac{\bar{L}_0^j}{\bar{L}_t^j}. \tag{2.40}$$

Therefore, if we are to model directly Libor rates as log-normally distributed satisfying the SDE

$$dL_t^i = L_t^i \left(\mu_t^i dt + \sigma_t^i dW_t^i \right), \tag{2.41}$$

then we need to derive the drift μ_t^i of each rate under $\mathbb{N}$. To this end, we apply Ito's formula to (2.40) to obtain the finite-variation part of M_{t,T_i} as

$$dM_t^i \doteq \sum_{j=1}^{i} \left(\sum_{k=1}^{i} \frac{\partial^2 M_t^i}{\partial \bar{L}_t^j \partial \bar{L}_t^k} L_t^j L_t^k \sigma_t^j \sigma_t^k dW_t^j dW_t^k + \frac{\partial M_t^i}{\partial \bar{L}_t^j} \mu_t^j L_t^j dt \right)$$

$$\doteq M_t^i \sum_{j=1}^{i} \frac{L_t^j}{\bar{L}_t^j} \left(\sum_{k=1}^{i} \frac{L_t^k}{\bar{L}_t^k} \sigma_t^j \sigma_t^k \rho_{i,j} - \mu_t^j \right) dt, \tag{2.42}$$

where $\doteq$ signifies that the two sides differ by a martingale term. Since M^i is an $\mathbb{N}$-martingale, the drift μ_t^i of the Libor rate $L_{t,[T_{i-1},T_i]}$ under $\mathbb{N}$ must therefore satisfy

$$\mu_t^i = \sigma_t^i \sum_{j=1}^{i} \frac{L_t^j}{\bar{L}_t^j} \rho_{i,j} \sigma_t^j. \tag{2.43}$$

While the BGM approach is appealing because it allows easy calibration to caplet prices, its application to the purpose of exposure computation is hindered by the fact that not all Libor rates are directly simulated, and therefore interpolation across tenors is required. More importantly, the simulated Libor rates are not directly obtainable from a common stochastic process, making any simulation computationally intensive.

2.4 Multiple Currencies and Foreign Exchange

We now turn to modelling of interest rates in a currency other than the reference one, leading naturally to modelling the exchange rate between the two currencies.

Table 2.2 Cross currency interest-rate products notation

$D_{t,T}$	T-bond in the local (domestic) currency
$\tilde{D}_{t,T}$	T-bond in the foreign currency
N_t	Numeraire in the local currency
$\tilde{N}_t$	Numeraire in the foreign currency
$M_{t,T}$	Martingale used to simulate T-bonds in the N_t (local) measure
$\tilde{M}_{t,T}$	Martingale used to simulate T-bonds in the $\tilde{N}_t$ (foreign) measure
χ_t	Spot FX rate from foreign to reference currency
Y_t	Change of measure from the reference measure $\mathbb{N}$ to the foreign measure $\tilde{\mathbb{N}}$

The basic tradable asset in the reference currency remains the bond, and we have modelled this in the previous section through the $\mathbb{N}$-martingale M:

$$M_{t,T} = \frac{D_{t,T}/D_{0,T}}{N_t/N_0}. \tag{2.44}$$

Whatever holds for the reference bond in the reference currency holds for the foreign one in the foreign currency, so in analogous manner we define another process (in terms of the foreign bond and foreign numeraire)

$$\tilde{M}_{t,T} = \frac{\tilde{D}_{t,T}/\tilde{D}_{0,T}}{\tilde{N}_t/\tilde{N}_0}, \tag{2.45}$$

which is now an $\tilde{\mathbb{N}}$-martingale, $\tilde{\mathbb{N}}$ being the measure that makes martingales out of all foreign tradable assets expressed in units of foreign numeraire.

Now, some relation is going to have to hold between the change-of-measure induced by M and $\tilde{M}$ and the exchange rate linking the foreign and reference currencies. Indeed, if $\chi \equiv (\chi_t)_{t \geq 0}$ is the FX process representing the value in reference currency of one foreign currency unit, then it is a standard no-arbitrage argument that the process $Y \equiv (Y_t)_{t \geq 0}$ defined by

$$Y_t = \frac{\chi_t}{\chi_0} \frac{\tilde{N}_t}{N_t} = \left. \frac{d\tilde{\mathbb{N}}}{d\mathbb{N}} \right|_{\mathscr{F}_t} \tag{2.46}$$

is the Radon-Nikodym derivative[7] for changing from the measure $\mathbb{N}$ to the measure $\tilde{\mathbb{N}}$. By this, Y_t is an $\mathbb{N}$-martingale.

Inserting $N_t = D_{0,t} M_{t,t}$ and $\tilde{N}_t = \tilde{D}_{0,t} \tilde{M}_{t,t}$ in (2.46), we get

$$Y_t = \frac{\chi_t}{\chi_0} \frac{D_{0,t}}{\tilde{D}_{0,t}} \frac{M_{t,t}}{\tilde{M}_{t,t}}. \tag{2.47}$$

---

[7]See Appendix B.

It turns out to be more convenient to express Y in terms of the basic FX forwards that one would observe in the market. To this end, define $\bar{F}_{t,T}$ to be

$$\bar{F}_{t,T} := F_{t,T}/F_{0,T} = \frac{\chi_t \tilde{D}_{t,T}/D_{t,T}}{\chi_0 \tilde{D}_{0,T}/D_{0,T}}, \qquad (2.48)$$

the time-t FX forward normalised by its time-zero value. In terms of this, Y has the simple representation

$$Y_t \tilde{M}_{t,T} = \bar{F}_{t,T} M_{t,T} \left(= \frac{\chi_t \tilde{D}_{t,T}/\chi_0 \tilde{D}_{0,T}}{N_t/N_0} \right) \qquad (2.49)$$

which is in fact an identity involving measure changes, namely

$$\left. \frac{d\tilde{\mathbb{N}}}{d\mathbb{N}} \right|_{\mathscr{F}_t} \left. \frac{d\tilde{\mathbb{N}}^T}{d\tilde{\mathbb{N}}} \right|_{\mathscr{F}_t} = \left. \frac{d\tilde{\mathbb{N}}^T}{d\mathbb{N}^T} \right|_{\mathscr{F}_t} \left. \frac{d\mathbb{N}^T}{d\mathbb{N}} \right|_{\mathscr{F}_t} \equiv \left. \frac{d\tilde{\mathbb{N}}^T}{d\mathbb{N}} \right|_{\mathscr{F}_t}, \qquad (2.50)$$

corresponding to changing from the reference numeraire measure to the foreign T-forward measure. It will prove useful to read off from (2.49) the facts that

$$Y_t \tilde{M}_{t,T}, \quad \bar{F}_{t,T} M_{t,T}, \quad Y_t, \quad \text{and} \quad M_{t,T} \qquad (2.51)$$

are all $\mathbb{N}$-martingales.

Intuitively, in (2.49) the volatility structure for $\bar{F}$ will relate to that of foreign exchange options,[8] while the volatility of M and $\tilde{M}$ arises from stochasticity of interest rates. The only component in (2.49) that is free to model is Y, and one would then obtain the FX process from this as

$$\chi_t = F_{0,t} \bar{F}_{t,t} = F_{0,t} Y_t \tilde{M}_{t,t}/M_{t,t}. \qquad (2.52)$$

The dynamics of Y have therefore to be chosen in such a way that market-observed FX option prices can be reproduced, given the dynamics of M and $\tilde{M}$ which would have already been calibrated to their respective interest-rate markets. In particular, the volatility structure for Y will depend on the volatilities and covariances of M, $\tilde{M}$ and $\bar{F}$.

In practice, doing this becomes very tedious especially when a large number of currencies are involved, so it is useful to look at a less exact but simpler approach.

---

[8]The volatility for the FX forwards $\bar{F}$ would be implied from market-observed option prices so that the price of an FX call struck at K, of maturity T, is

$$C(K,T) = \mathbb{E}^{\mathbb{N}} \left[N_T^{-1} (\chi_T - K)^+ \right]$$

$$= D_{0,T} F_{0,T} \tilde{\mathbb{N}}^T [\chi_T^{-1} \leq K^{-1}] - D_{0,T} K \mathbb{N}^T [\chi_T \geq K],$$

where we recall $F_{0,T} D_{0,T} = \chi_0 \tilde{D}_{0,T}$ and where we have used the identities (2.50) to switch from one measure to another. Note in the last equality that $\bar{F}$ (resp. $\bar{F}^{-1}$) is a martingale in the forward measure $\mathbb{N}^T$ (resp. $\tilde{\mathbb{N}}^T$).

We go about this by modelling $\bar{F}$ (and therefore χ) *as if it were independent of the interest-rate martingales M and $\tilde{M}$*. In effect, we mimic (2.49) by

$$Y_t \tilde{M}_{t,T} = \frac{\chi_t \tilde{D}_{t,T} / \chi_0 \tilde{D}_{0,T}}{N_t / N_0} \approx \hat{F}_{t,T} M_{t,T} \tag{2.53}$$

with $\hat{F}$ independent of M and $\tilde{M}$ but having the same marginal distributions as $\bar{F}$. The reason this helps is that the market prices for FX options tell us directly what the volatility structure for $\bar{F}$ must be. The immediate consequence of this independence assumption[9] is that

$$\hat{F}_{t,T} \text{ is an } \mathbb{N}\text{-martingale.}$$

The cost of this assumption is of course that while $M\bar{F}/\tilde{M}$ and $M\bar{F}$ are $\mathbb{N}$-martingales, $M\hat{F}/\tilde{M}$ and $M\hat{F}$ are not; to mitigate this we impose a drift correction on $\tilde{M}_{t,T}$ so as to ensure at least that their *expected* values stay constant (and equal to one), that is

$$\mathbb{E}\left[M_{t,T} \hat{F}_{t,T} / \tilde{M}_{t,T} \right] = 1 \left(= \mathbb{E}\left[M_{t,T} \bar{F}_{t,T} / \tilde{M}_{t,T} \right] = \mathbb{E}(Y_t) \right)$$

$$\mathbb{E}\left[M_{t,T} \hat{F}_{t,T} \right] = 1 \left(= \mathbb{E}\left[\frac{\chi_t \tilde{D}_{t,T} / \chi_0 \tilde{D}_{0,T}}{N_t / N_0} \right] \right). \tag{2.54}$$

Using the independence between $\hat{F}$ and the pair $(M, \tilde{M})$, it can be seen that

$$\mathbb{E}\left[\tilde{M}_{t,T} \right] = 1 - \text{Cov}\left(M_{t,t} \tilde{M}_{t,t}^{-1}, \tilde{M}_{t,T} \right). \tag{2.55}$$

We re-iterate here that the above expectations are taken on time-zero information, that is, in the filtration $\mathscr{F}_0$. Thus, the drift correction that we derive for $\tilde{M}$ is the *expected* drift seen at time zero. Attempting similar calculations for arbitrary $\mathscr{F}_t$ fail because the processes $\hat{F}M$ and $\hat{F}M/\tilde{M}$ are not bona-fide martingales. This arises as a direct consequence of the independence assumption for $\hat{F}$; this independence is of course inconsistent with the fact that the conditional expectations of the FX forwards do depend on observed bond prices.

   In practice, the input to the model is the correlation between the drivers of M and $\tilde{M}$ in the $\mathbb{N}$-measure, and in concrete examples, the covariances above need to be expressed in terms of this correlation. In particular, when M and $\tilde{M}$ are modelled with deterministic volatilities (see (2.15)), the condition (2.55) is expressed by saying that the Brownian Motion $\tilde{X}_t$ driving the martingale $\tilde{M}_{t,T}$ (see (2.24)) has, in the $\mathbb{N}$-measure, the drift

$$\alpha_{t,T} := \tilde{\sigma}_{t,t} - \rho \sigma_{t,t} \equiv \alpha(t), \tag{2.56}$$

---

[9] ...it also implies that while rates in different currencies can have co-dependence, all other asset classes are independent of interest rates. ...

where ρ is the instantaneous correlation between the Brownians $\tilde{X}$ and X that drive the martingale M and $\tilde{M}$, respectively. Note that (2.56) is a function only of t, so that the Brownian Motion $\tilde{X}_t$ looks like

$$B_t + \int_0^t \alpha(u)du, \qquad (2.57)$$

in terms of an $\mathbb{N}$-Brownian Motion B having correlation ρ with X.

2.5 Inflation

The basic inflation product, an inflation-linked bond, is designed to preserve the purchasing power of money in some given currency. Modelling of inflation products is generally similar to modelling of foreign exchange since for each currency, one can consider the nominal and real rates to be like a conventional (local and foreign, respectively) currency pair. Inflation yield curves (arising from inflation-linked bonds) and volatility term structures (deduced from prices of inflation-linked options) serve to calibrate the inflation model, in the same way that foreign currency yield curves and FX options allow calibration of foreign exchange models.

In concrete terms, we let the exchange rate χ represent the inflation index denominated in the reference currency, with $\tilde{M}$ being calibrated to the volatility of real rates. For example, if the reference currency is GBP, the UK RPI inflation rate (denominated in GBP) would be modelled as some process χ, as if the RPI index were a foreign currency.

2.6 Equity

Equity can be tackled within our framework by treating it as an asset foreign to the reference currency, similar to what we do for foreign currencies. In this way, one (i) lets the 'exchange rate' χ represent the value in its denomination currency of one unit of stock and (ii) uses $\tilde{M}$ to control the volatility of the stock dividend yields.

In more detail, recall the identity (2.49)

$$Y_t \tilde{M}_{t,T} = \bar{F}_{t,T} M_{t,T} = \frac{\chi_t \tilde{D}_{t,T}/\chi_0 \tilde{D}_{0,T}}{N_t/N_0}. \qquad (2.58)$$

The third term here represents the tradable asset for the foreign currency, namely the price of the foreign bond expressed in the reference currency. Now consider a stock denominated in a foreign currency. The stock forward price, $S_{t,T}$, discounted by the foreign bond and expressed in units of the foreign numeraire, is the relevant traded asset for the stock and has the representation as the $\tilde{\mathbb{N}}$-martingale

$$\tilde{Y}_t^{(S)} M_{t,T}^{(S)} := \bar{S}_{t,T} \tilde{M}_{t,T} = \frac{S_{t,T} \tilde{D}_{t,T}/S_{0,T} \tilde{D}_{0,T}}{\tilde{N}_t/\tilde{N}_0}, \qquad (2.59)$$

with S (respectively, $\bar{S}$) being the stock forwards (respectively, forwards normalised by their initial values), denominated in their own currency, and with $\tilde{Y}^{(S)}$ and $\tilde{M}$ being $\tilde{\mathbb{N}}$-martingales.

The martingale $M^{(S)}$ will control the volatility of the stock dividends, so the element in (2.59) that is free to model is now $\tilde{Y}^{(S)}$, exactly as what happens for Y when modelling the FX rate. What remains is to ask what the $\tilde{\mathbb{N}}$-martingale $\tilde{Y}^{(S)}$ looks like in $\mathbb{N}$.

But because $Y_t = d\tilde{\mathbb{N}}/d\mathbb{N}|_{\mathscr{F}_t}$, and $\tilde{Y}^{(S)}$ is an $\tilde{\mathbb{N}}$-martingale, we have that $Y\tilde{Y}^{(S)}$ is an $\mathbb{N}$-martingale. Thus, for each $s < t$,

$$\tilde{Y}_s^{(S)} Y_s = \mathbb{E}\left[\tilde{Y}_t^{(S)} Y_t \Big| \mathscr{F}_s\right]$$

$$= Y_s \mathbb{E}\left[\tilde{Y}_t^{(S)} \Big| \mathscr{F}_s\right] + \mathrm{Cov}\left[Y_t, \tilde{Y}_t^{(S)} \Big| \mathscr{F}_s\right], \tag{2.60}$$

where we have used the $\mathbb{N}$-martingale property of Y. Dividing out Y_s we get that

$$\tilde{Y}_s^{(S)} = \mathbb{E}\left[\tilde{Y}_t^{(S)} \Big| \mathscr{F}_s\right] + Y_s^{-1}\mathrm{Cov}\left[Y_t, \tilde{Y}_t^{(S)} \Big| \mathscr{F}_s\right]. \tag{2.61}$$

The second term on the right here is commonly referred to as a quanto adjustment.

In the case where Y and $Y^{(S)}$ are both log-normally distributed with deterministic volatility, the equation above will result in a drift term for $Y^{(S)}$ of the form

$$\mu_t^{(S)} = -\rho_t \sigma_t \sigma_t^{(S)}, \tag{2.62}$$

where ρ_t, σ_t and $\sigma_t^{(S)}$ are respectively the instantaneous correlation linking Y and $\tilde{Y}^{(S)}$, the instantaneous volatility of Y and the instantaneous volatility of $\tilde{Y}^{(S)}$.

Table 2.3 Notation for equity products denominated in a foreign currency

$S_{t,T}$	Forward stock price in the stock's currency
$\bar{S}_{t,T}$	Normalised stock forward price, $S_{t,T}/S_{0,T}$
$\tilde{Y}^{(S)}$	Change of measure $d\mathbb{N}^{(S)}/d\tilde{\mathbb{N}}$

2.7 Credit

For credit products, exposure is driven by (i) the likelihood and occurrence of defaults, (ii) the dynamics of credit spreads, and (iii) the inter-dependence between defaults of different entities. Often the fundamental quantities chosen to be modelled are the credit spreads, as they are directly observable in the market (see for example the books by Bielecki & Rutkowski [14], Duffie & Singleton [39], Lando [72] or Schönbucher [95] for a discussion about credit models). We opt for a model whereby stochastic default probabilities are simulated directly. As we will see this allows us

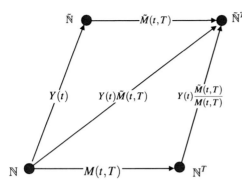

Measure	Numeraire
$\mathbb{N}$	N_t
$\mathbb{N}^T$	$\dfrac{D_{t,T}}{D_{0,T}}$
$\tilde{\mathbb{N}}$	$\dfrac{\chi_t \tilde{N}_t}{\chi_0}$
$\tilde{\mathbb{N}}^T$	$\dfrac{\tilde{D}_{t,T}}{\tilde{D}_{0,T}} \dfrac{\chi_t}{\chi_0}$

Fig. 2.1 Change of measures between the local measure $\mathbb{N}$, the T-forward local measure $\mathbb{N}^T$, the foreign measure $\tilde{\mathbb{N}}$ and the T-forward foreign measure $\tilde{\mathbb{N}}^T$

to work in a framework similar to what we have introduced in the previous sections. Section 2.7.1 below describes how the par CDS spreads observed in the market can be used to obtain initial default probabilities. After that, we propose a method for evolving default probabilities of single entities, simulating default times, and modelling inter-dependence of credit spreads and default times of different entities.

The model as we present it imposes a Gaussian dependence structure, which is used to achieve dependence between default probabilities and simulated default times of any pair of reference entities. Such Gaussian models are widely used in practice, even if it is well known that the Gaussian copula does not build dependence in the tail of the distribution. Other, more realistic, ways of introducing dependence can be used (for a discussion of other types of copula see, for example, [78]), but the mathematics is bound to become much more tedious.

2.7.1 Default Probabilities from par CDS Spreads

At any time t, the observed CDS curve consists of a set of spreads, s_{t,T_i} for a set of maturities T_i, $i = 1, 2, \ldots, n$. The spreads indicate, for a given entity which is not yet in default at time t, the market view on the propensity of that entity to default in the future. By writing down the value of the CDS contract,[10] we can express default probabilities for the horizons T_i in terms of the set of spreads s_{t,T_i} observed at time t.

---

[10] A CDS (credit default swap) is a product entitling its holder to receive, at time of default and in return for a bond issued by an entity that has just defaulted, the face value of that bond. A common convention when pricing CDSs is to assume that the value of a defaulted bond is a fraction R of its face value. Under such a convention, the net value of the CDS at the time of default is then $\bar{R} := (1 - R)$ per unit of face value. We will describe CDSs in detail in Chap. 10.

Table 2.4 Notation for credit products

$\tau, \tau^{(i)}$	Generic default time, default time for reference name i
$q_{t,T} \equiv 1 - p_{t,T}$	Probability of surviving beyond T, conditional on t-information and survival until t
$\bar{M}_{t,T}$	The $\mathbb{N}$ martingale $\mathbb{E}[\mathbb{1}_{\tau > T} \mid \mathscr{F}_t]$

To see this, consider a CDS contract of maturity T_n, for which the observed spread is s_{t,T_n}, and write $T_1, \ldots, T_n$ for the fee payment dates of the CDS. The distance between payment dates is $\alpha_i = T_i - T_{i-1}$, with $t = T_0 < T_1$. Set

$$p_{t,T} := \mathbb{N}\left(\tau \in (t, T) \mid \mathscr{F}_t, \tau > t\right) \qquad (2.63)$$

to be the probability in the time-t filtration that the default time τ of the entity underlying the CDS lies in $(t, T]$ (given survival until t). Then the fee leg of the CDS has value

$$A_t := s_{t,T_n} \sum_{i=1}^{n} D_{t,T_i} \alpha_i (1 - p_{t,T_i}). \qquad (2.64)$$

Similarly, if $\bar{R} \equiv 1 - R$ represents the payment per unit notional made by the CDS upon default, the value of protection offered by the CDS is given by

$$B_t := \bar{R} \sum_{i=1}^{n} D_{t,T_i} (p_{t,T_i} - p_{t,T_{i-1}}), \qquad (2.65)$$

where we have assumed that default payments are made at time-points in a discretized grid. The observed (fair) CDS spread s_{t,T_n} is that which makes A_t and B_t equal. That is to say, the unknown probabilities $p_{t,T_1}, \ldots, p_{t,T_n}$ satisfy

$$\sum_{i=1}^{n} D_{t,T_i} \left\{\alpha_i s_{t,T_n} + \bar{R} p_{i-1} - (\alpha_i s_{t,T_n} + \bar{R}) p_i\right\} = 0, \qquad (2.66)$$

where we have abbreviated $p_j \equiv p_{t,T_j}$.

For $n = 1$, (2.66) is an expression in p_1 and $p_0 \equiv 0$, and therefore gives us the value of p_1 as

$$p_1 = \left[\alpha_1 s_{t,T_1}\right] / \left[\alpha_1 s_{t,T_1} + \bar{R}\right]. \qquad (2.67)$$

Similarly, given values of $p_1, \ldots, p_{j-1}$, we obtain the value of p_j in terms of $p_i, i = 1, \ldots, j-1$ and s_{t,T_j}. Explicitly,

$$p_j = \frac{D_{t,T_j}^{-1}\left[\sum_{i=1}^{j-1} D_{t,T_i}\{\alpha_i s_{t,T_n} + \bar{R} p_{i-1} - (\alpha_i s_{t,T_n} + \bar{R}) p_i\}\right]}{\alpha_j s_{t,T_j} + \bar{R}}$$

$$+ \frac{\alpha_j s_{t,T_j} + \bar{R} p_{j-1}}{\alpha_j s_{t,T_j} + \bar{R}}, \qquad j = 2, 3, \ldots, n. \qquad (2.68)$$

The inductive recipe (2.68) gives us the per-period default probabilities p_{t,T_i}, which at time t are consistent with CDS market prices observed as the spreads s_{t,T_i}. In particular, at $t = 0$, we can obtain probabilities p_{0,T_i} which are consistent with time-zero observed CDS spreads.

2.7.2 Stochastic Default Probabilities

The time-zero default probabilities (2.68) serve as the starting point for the simulation of stochastic default probabilities for a given entity.

Our modelling hinges on specifying dynamics for the $\mathbb{N}$-martingale

$$\bar{M}_{t,T} = \mathbb{E}\left[\mathbb{1}_{\tau > T} \mid \mathscr{F}_t\right]. \tag{2.69}$$

Here, expectation is taken in the martingale measure corresponding to numeraire N, where we will assume that credit quantities and interest rates are independent.[11]

The initial values $\bar{M}_{0,T}$ in (2.69) are chosen to replicate exactly the initial term structure of survival probabilities, namely

$$\bar{M}_{0,T} = (1 - p_{0,T}) \equiv q_{0,T}. \tag{2.70}$$

The filtration $\mathscr{F} = (\mathscr{F}_t)_{t \geq 0}$ contains information about the underlying drivers of the economy, but *not* about actual defaults. In this filtration, the martingale M is related to survival probabilities via the intensity process, $\lambda = (\lambda_t)_{t \geq 0}$, for τ, that is,

$$\begin{aligned}
\bar{M}_{t,T} &= \mathbb{E}\left[\mathbb{1}_{\tau > T} \mid \mathscr{F}_t\right] \\
&= \exp\left(-\int_0^t \lambda_u du\right) \mathbb{E}\left[\exp\left(-\int_t^T \lambda_u du\right) \mid \mathscr{F}_t\right] \\
&=: \bar{M}_{t,t} q_{t,T},
\end{aligned} \tag{2.71}$$

where the last equality serves to define $q_{t,T}$ as the probability that conditional on having survived until time t, and conditional on the information in $\mathscr{F}_t$, the reference name does not default before T.

From (2.71), default probabilities $q_{t,T}$ can be obtained from values of martingales $\bar{M}_{\cdot,T}$. This approach is in a sense akin to those followed by Sidenius [99] and Bennani [12].

The form of $\bar{M}_{t,t}$ also points to how one can simulate values of the default time for any given reference name:

(i) draw a random uniform U;
(ii) set τ to be the least member of the set $\{t \geq 0 : \bar{M}_{t,t} \leq U\}$.

---

[11]The implication of this is that $\bar{M}_{t,T}$ is a martingale under any rate-based numeraire measure.

Looking at the recipe above, correlation between different reference names can be introduced by

(i) imposing dependence between $U^{(i)}$ and $U^{(j)}$ for any i, j, and/or
(ii) setting up diffusion dynamics for the martingales $\bar{M}_{t,T}^{(i)}$ and $\bar{M}_{t,T}^{(j)}$ and then allowing the driving Brownian Motions to be correlated.

The question of how to introduce dependence between default times has been approached in several ways in the literature, as any standard book on credit derivatives will reveal (see for example Lando [72] or Schoenbucher [95]). We approach the problem using a copula method. One of several copulas can be used, each type of copula building dependence in a different way (see, for example Madan et al. [77]). In our exposition below, we use the example of the well-known Gaussian copula, for which dependence is equivalent to a linear correlation parameter.

Thus, we specify for each reference name i,

$$U^{(i)} = 1 - \Phi\left(\boldsymbol{\rho}^{(i)} \cdot \mathbf{M} + \bar{\rho}^{(i)} M^{(i)}\right)$$

$$dW^{(i)} = \boldsymbol{\eta}^{(i)} \cdot d\mathbf{Z} + \bar{\eta}^{(i)} dZ^{(i)}.$$

(2.72)

Here, $\mathbf{M}$ and $M^{(i)}$ are standard normally distributed random variables (the former of dimension possibly larger than one) driving dependence between default times. Similarly, $\mathbf{Z}$ and $Z^{(i)}$ are standard Brownian Motions (with the former again of dimension possibly larger than one) driving the stochasticity of survival probabilities. The reals $\bar{\rho}^{(i)} = \sqrt{1 - \boldsymbol{\rho}^{(i)} \cdot \boldsymbol{\rho}^{(i)}}$ and $\bar{\eta}^{(i)} = \sqrt{1 - \boldsymbol{\eta}^{(i)} \cdot \boldsymbol{\eta}^{(i)}}$ ensure that $U^{(i)}$ and $dW^{(i)}$ have uniform and normal distributions, respectively.

For each pair (i, j), $\mathbf{Z}$, $Z^{(i)}$, $Z^{(j)}$, $dW^{(i)}$ and $dW^{(j)}$ are all independent; expressions (2.72) build up dependence between $U^{(i)}$, $U^{(j)}$, and between $dW^{(i)}$, $dW^{(j)}$, because defaults of all entities depend on $\mathbf{Z}$ and on $\mathbf{M}$. Elements in $\mathbf{M}$ and $\mathbf{Z}$ are thought of as market factors impacting defaults and spread evolutions for different entities, while $M^{(i)}$ and $Z^{(i)}$ affect only the spread evolution and default time of reference name $^{(i)}$.

We will see in Chap. 3 the specific form the inter-name dependence takes for a particular choice of dynamics for the martingale $\bar{M}$.

2.7.3 Loss Simulation

For credit products that depend on defaults of several entities, the simulation of defaults for individual names implies, assuming an appropriate dependence structure has been imposed, a corresponding simulation of losses suffered by a chosen portfolio of names.

Consider an investor who holds several corporate bonds $\bar{B}^{(i)}$, $i = 1, 2, \ldots, n$. The loss l_i incurred by bond i upon default is

$$l_i = (1 - R_i) A_i \equiv \bar{R}_i A_i,$$

(2.73)

where A_i is the nominal amount on bond i and R_i is the fraction of face value that is retained by the bond upon default. Given this, the fractional total loss suffered by the portfolio of bonds in the time interval $[0, t]$ is defined to be

$$L_t = \sum_{i=1}^{n} \bar{R}_i A_i \mathbb{1}_{\{\tau^{(i)} \leq t\}} \bigg/ \sum_{i=1}^{n} A_i. \qquad (2.74)$$

The law of L_t depends of course on the dependence between default times of different entities. The typical observable quantity that provides information on such dependence is the market price of a CDO tranche, which is a product whose payoff at time t is of the form

$$\Pi_t = \left(\sum_{i=1}^{n} L_t - k_a \right)^+ - \left(\sum_{i=1}^{n} L_t - k_d \right)^+, \qquad (2.75)$$

where k_a (respectively, k_d), satisfying

$$0 \leq k_a \leq k_d \leq 1,$$

are referred to as the attachment (respectively, detachment) point.[12]

We will see in Chap. 3 how market quotes for different tranches can be used to calibrate the dependence coefficients ρ and η appearing in (2.72).

---

[12]CDO (Collateral Debt Obligation) products will be described in details in Chap. 10.

Chapter 3
Simulation Models

In Chap. 2 we defined a general framework to enable estimation of counterparty exposure for different product classes. Throughout, we highlighted the importance of being able to simulate price processes of different asset classes simultaneously and in consistent fashion. This was accomplished by simulating a martingale process for each asset class. By doing so, the models fit time-zero forward curves by construction, so that calibration involves only choosing the volatility structure for the martingale pertaining to each asset class.

In this chapter we focus on specific choices of models for different asset classes, discussing how they can be implemented and calibrated within our framework.

3.1 Interest-Rate Models

For relatively simple interest-rate products, arbitrage-free models with deterministic volatility have dynamics rich enough to reproduce simulated price distributions with the correct properties. In this section we describe in detail, within the framework defined in Chap. 2, the model with separable volatility structure introduced in (2.23), and we show in particular how the familiar Hull-White model is a special case of such a separable specification.

For ease of exposition, we will mostly refer to a separable model driven by a single Brownian Motion. Products such as steepeners, however, which depend on different points of the yield curve, may require models with more than one stochastic driver or a richer volatility structure.[1]

---

[1] There is a vast literature on interest-rate models. The reader can refer to the following books, Brigo & Mercurio [18], Cairns [21], Filipovic [44], Hunt & Kennedy [64], Pelsser [85], or Rebonato [89] for more details.

G. Cesari et al., *Modelling, Pricing, and Hedging Counterparty Credit Exposure,*
Springer Finance, DOI 10.1007/978-3-642-04454-0_3,
© Springer-Verlag Berlin Heidelberg 2009

3.1.1 Separable Volatility

In Sect. 2.3 we expressed all bond prices and the numeraire in terms of a family of
$\mathbb{N}$-martingales $M_{.,T}$. In turn, for deterministic volatilities, (2.15) gives the integral
form of $M_{t,T}$ as

$$M_{t,T} = M_{0,T} \exp\left(\int_0^t \sigma_{u,T} dW_{u,T} - \frac{1}{2} \Sigma_{t,T}^2 t \right), \tag{3.1}$$

in terms of $\mathbb{N}$-Brownian Motions $W_T \equiv (W_{t,T})_{0 \le t \le T}$ and volatility functions $\sigma \equiv$
$(\sigma_{t,T})$. What we study here is the special case, already alluded to in (2.23), where
the dependence of the volatility function on t and T can be separated into two terms.
That is, we look at cases where the SDE for M can be written as (see also Sect. 2.3.1)

$$dM_{t,T} = M_{t,T} \mathbf{g}_T \cdot (\mathbf{f}_t \mathbf{R} d\mathbf{W}_t), \tag{3.2}$$

with $\mathbf{W}$ a Brownian Motion in $\mathbb{R}^n$, $\mathbf{g}_T$ a deterministic vector in $\mathbb{R}^n$, and with $\mathbf{R}^\mathsf{T} \mathbf{R}$
being a positive semi-definite matrix such that the i'th component of $\mathbf{RW}$ is a stan-
dard Brownian Motion, which then gets time-changed by the (i, i)'th entry in the
diagonal $(n \times n)$ matrix $\mathbf{f}_t$. Thus, by defining a new, time-changed process (also in
$\mathbb{R}^n$)

$$\mathbf{X_t} = \int_0^t \mathbf{f}_u \mathbf{R} d\mathbf{W}_u, \tag{3.3}$$

the expression for $M_{t,T}$ can be written as

$$M_{t,T} = M_{0,t} \exp\left(\mathbf{g}_T \cdot \mathbf{X}_t - \frac{1}{2} \Sigma_{t,T}^2 t \right). \tag{3.4}$$

The term $\Sigma_{t,T}^2 t$ is the variance of $\mathbf{g}_T \cdot \mathbf{X}_t$, that is,

$$\Sigma_{t,T}^2 := t^{-1} \mathrm{Var}\,(\mathbf{g}_T \cdot \mathbf{X}_t) = \mathbf{g}_T \cdot \left(\int_0^t (\mathbf{f}_u \mathbf{R})^\mathsf{T} (\mathbf{f}_u \mathbf{R}) du \right) \mathbf{g}_T$$

$$=: \mathbf{g}_T \cdot \mathbf{F}_t^2 \mathbf{g}_T, \tag{3.5}$$

serving to define the matrix $\mathbf{F}$. While we have written the model in general for n
factors, to simplify the notation we will mostly restrict ourselves to the one-factor
case in what follows; the vector $\mathbf{g}_T$ and the matrix $\mathbf{f}_t$ are then real-valued functions
of T and t, respectively.

In general $\Sigma_{t,T}$ will be characterised by a number of parameters, and in a simulation framework, it is necessary to first calibrate these to instruments whose values
are observable in the market. Thus, for instance, by writing the time-t price of a
T-bond in terms of the martingale M, we get that

$$D_{t,T} = \frac{D_{0,T}}{D_{0,t}} \frac{M_{t,T}}{M_{t,t}}$$

$$= \frac{D_{0,T}}{D_{0,t}} \frac{\exp\left(g_T X_t - \frac{1}{2}\Sigma_{t,T}^2 t\right)}{\exp\left(g_t X_t - \frac{1}{2}\Sigma_{t,t}^2 t\right)}, \tag{3.6}$$

where in this one-factor case the variance is simply

$$t\Sigma_{t,T}^2 = t F_t^2 g_T^2. \tag{3.7}$$

The functional dependence of F and g on T and t may now be chosen to have the desired behaviour for the volatility of the T-bond. For instance, taking $f_t \equiv 1$ and

$$g_T = a\left(1 - e^{-\kappa T}\right) \tag{3.8}$$

results in bond volatilities

$$(g_T - g_t)F_t = ae^{-\kappa t}\left(1 - e^{-\kappa \tau}\right) \quad (\tau \equiv T - t), \tag{3.9}$$

which decrease with with time t but increase with tenor τ. In other words, bond options (or, equivalently, *caplets*; see below) priced with this model would exhibit implied volatilities that decrease with expiry but increase with tenor.

One will need to choose the number of model parameters so as to strike the right balance between having a model that is parsimonious enough and having a fit to market prices that is good enough. It is also important to note that calibration can influence materially the counterparty risk profile. In the next sections we will show how to calibrate the model, using caps, floors, and swaptions.

3.1.1.1 Calibration to Caps

A cap is a market instrument that pays, at pre-specified points in time, $T_1 < \cdots < T_n$, the amount by which the observed Libor rate exceeds a given level, K, referred to as the cap strike. Thus, the time-zero value, denoted $\mathscr{C}_{0,T_k}(K)$, of the k'th payment made by the cap is

$$\mathscr{C}_{0,T_k}(K) = D_{0,T_k}\mathbb{E}^{\mathbb{N}^{T_k}}\left[\alpha_k\left(L_{T_{k-1},[T_{k-1},T_k]} - K\right)^+\right]; \tag{3.10}$$

the payoff above is referred to as a *caplet*. T_k indicates the maturity of the caplet, and $\mathbb{N}^{T_k}$ the T_k-forward measure. On the right, the payoff is a call option on the Libor rate, fixed at T_{k-1}, for the period $[T_{k-1}, T_k]$ of length $\alpha_k = T_k - T_{k-1}$. Changing measure from $\mathbb{N}^{T_k}$ to $\mathbb{N}$ in (3.10) using (2.7), we get that

$$\mathscr{C}_{0,T_k}(K) = \alpha_k D_{0,T_k}\mathbb{E}\left[M_{T_{k-1},T_k}\left(L_{T_{k-1},[T_{k-1},T_k]} - K\right)^+\right]$$

$$= \alpha_k D_{0,T_k}\mathbb{E}\left[\left(\bar{L}_k M_{T_{k-1},T_{k-1}} - \bar{K}M_{T_{k-1},T_k}\right)^+\right] \quad \text{(from (2.19))}$$

$$= \alpha_k D_{0,T_k} \bar{L}_k \mathbb{N}^{T_{k-1}} \left[\frac{M_{T_{k-1},T_k}}{M_{T_{k-1},T_{k-1}}} \leq \frac{\bar{L}_k}{\bar{K}} \right]$$

$$- \alpha_k D_{0,T_k} \bar{K} \mathbb{N}^{T_k} \left[\frac{M_{T_{k-1},T_{k-1}}}{M_{T_{k-1},T_k}} \geq \frac{\bar{K}}{\bar{L}_k} \right], \tag{3.11}$$

where $\bar{L}_k = L_{0,[T_{k-1},T_k]} + \alpha_k^{-1}$ and $\bar{K} = K + \alpha_k^{-1}$. Define

$$\bar{M}_t^k = \frac{M_{t,T_{k-1}}}{M_{t,T_k}}. \tag{3.12}$$

Since $\bar{M}^k$ and $(\bar{M}^k)^{-1}$ are respectively martingales in $\mathbb{N}^{T_k}$ and $\mathbb{N}^{T_{k-1}}$, evaluating the probabilities in (3.11) involves knowing the volatility structure of $\bar{M}^k$. In particular, in the case of a separable volatility structure, we can write these probabilities as

$$\mathbb{N}^{T_{k-1}} \left[\left(\bar{M}_{T_{k-1}}^k \right)^{-1} \leq \frac{\bar{L}_k}{\bar{K}} \right] = \Phi\left(d_k \right), \quad \text{and}$$

$$\mathbb{N}^{T_k} \left[\bar{M}_{T_{k-1}}^k \geq \frac{\bar{K}}{\bar{L}_k} \right] = \Phi\left(d_k - s_{T_{k-1},T_k} \sqrt{T_{k-1}} \right), \tag{3.13}$$

where

$$d_k := \frac{\ln\left(\bar{L}_k / \bar{K} \right)}{s_{T_{k-1},T_k} \sqrt{T_{k-1}}} + \frac{s_{T_{k-1},T_k} \sqrt{T_{k-1}}}{2}. \tag{3.14}$$

The variance of $\bar{M}_{T_{k-1}}^k$ is

$$s_{T_{k-1},T_k}^2 = \left(\mathbf{g}_{T_{k-1}} - \mathbf{g}_{T_k} \right) \cdot \mathbf{F}_{T_{k-1}}^2 \left(\mathbf{g}_{T_{k-1}} - \mathbf{g}_{T_k} \right), \tag{3.15}$$

where $\mathbf{F}_{T_{k-1}}$ is the diagonal matrix defined in (3.5).

3.1.1.2 Calibration to Swaptions

An interest-rate swaption is an instrument that gives the holder the right to enter into a fixed-for-floating interest-rate swap. Such instruments can be valued in closed-form in the case of separable volatility one-factor models, and can therefore be used for calibration to observed market prices.

Consider for concreteness a swaption which can be exercised at time T to enter into a swap where the holder pays a fixed annualised coupon of K and receives the Libor rate of interest. Write, again, $T_1 < \cdots < T_n$ for the times when the swap payments are exchanged, and define $(T_k - T_{k-1}) = \alpha_k$ for all $k = 1, 2, \ldots, n$. Assume that $T = T_0 = T_1 - \alpha_1$, so that the swaption exercise time coincides with the time of the first fixing on the floating rate component of the underlying swap. Then the time-zero value, $\mathscr{S}_{0,T,T_n}(K)$, of the swaption is given by

$$\mathscr{S}_{0,T,T_n}(K) = D_{0,T}\,\mathbb{E}^{\mathbb{N}^T}\left[\left(1 - D_{T,T_n} - K\sum_{k=1}^{n}\alpha_k D_{T,T_k}\right)^+\right]$$

$$= \mathbb{E}^{\mathbb{N}^T}\left[\left(D_{0,T} - \sum_{k=1}^{n}a_k D_{0,T_k}\frac{M_{T,T_k}}{M_{T,T}}\right)^+\right], \qquad (3.16)$$

where

$$a_k = \begin{cases} K\alpha_k, & k = 1, 2, \ldots, n-1, \\ 1 + K\alpha_k, & k = n. \end{cases} \qquad (3.17)$$

In the special case of a one-factor model, (3.16) can be expressed as a sum of put options of different strikes. Indeed, if for any T, $\frac{M_{t,T}}{M_{t,t}}$ is a deterministic, monotonic function of a single stochastic process X_t, then we can define

$$h_i(X_t) := M_{t,T_i}/M_{t,T}$$

$$h(X_t) := D_{0,T} - \sum_{i=1}^{n}a_i D_{0,T_i}\frac{M_{t,T_i}}{M_{t,T}} = D_{0,T} - \sum_{i=1}^{n}a_i D_{0,T_i}h_i(X_t) \qquad (3.18)$$

with h and the h_i's being bijective.

Therefore, there exists a family of strikes K_i^*, $i = 1, 2, \ldots, n$, such that

$$\left(D_{0,T} - \sum_{i=1}^{n}a_i D_{0,T_i}\frac{M_{T,T_i}}{M_{T,T}}\right)^+ = M_{T,T}^{-1}\sum_{i=1}^{n}a_i D_{0,T_i}(K_i^* M_{T,T} - M_{T,T_i})^+. \quad (3.19)$$

In fact, the above is satisfied by having

$$K_i^* = h_i\left(h^{-1}(0)\right). \qquad (3.20)$$

Once the strikes K_i^* have been found using a root-searching algorithm, pricing the swaption is reduced to pricing put options of the form

$$\mathbb{E}^{\mathbb{N}^T}\left[\left(K_i^* - \frac{M_{T,T_i}}{M_{T,T}}\right)^+\right]$$

$$= K_i^*\Phi(d) - \Phi\left(d - s_{T,T_i}\sqrt{T}\right), \qquad (3.21)$$

where

$$d := \frac{\ln K_i^*}{s_{T,T_i}\sqrt{T}} + \frac{s_{T,T_i}\sqrt{T}}{2}, \qquad (3.22)$$

$$s_{T,T_i}^2 = F_T^2\left(g_T - g_{T_i}\right)^2. \qquad (3.23)$$

3.1.2 Example: Hull-White (Extended Vasicek)

We analysed in the previous section how to calibrate with caps and swaptions a generic model specified in terms of separable volatility structure. We consider here a specific example of separable volatility model, the Hull-White model.

The Hull-White model introduced in [63] is a short rate model with deterministic volatility, where the short rate r satisfies the SDE

$$dr_t = (\theta_t - ar_t)dt + \sigma dW_t, \tag{3.24}$$

with speed of mean reversion a and volatility σ (see also Baxter & Rennie [10], Brigo & Mercurio [18], or Hull [62] for further details). Knowing the instantaneous forward rate $f(0, t)$ at time zero for different values of t (that is, knowing the current yield curve) allows determination of the unknown θ_t in (3.24). Note also that (3.24) prescribes that all bonds in the economy will depend on the same source of randomness W. We can write for each $t \geq 0$,

$$r_t = f(0, t) + \frac{\sigma^2}{2a^2}(1 - e^{-at})^2 + \sigma \int_0^t e^{-a(t-u)}dW_u, \tag{3.25}$$

in terms of $f(0, t)$. From this it is immediate that for each t, the short rate r_t is a normal random variable with variance

$$\sigma^2 \frac{1 - e^{-2at}}{2a}. \tag{3.26}$$

Moreover, note that from (3.25), the Hull-White numeraire process satisfies

$$\begin{aligned} N_t^{-1} &= \exp\left\{-\int_0^t f(0, u)du - \int_0^t \frac{\sigma^2}{2a^2}(1 - e^{-au})^2 du \right. \\ &\quad \left. - \int_0^t \sigma \int_0^u e^{-a(u-s)}dW_s du\right\} \\ &= D_{0,t} \exp\left\{-\int_0^t \frac{\sigma^2}{2a^2}(1 - e^{-au})^2 du - \int_0^t \sigma \int_0^u e^{-a(u-s)}dW_s du\right\} \\ &= D_{0,t} \exp\left\{-\int_0^t \frac{\sigma^2}{2a^2}(1 - e^{-au})^2 du - \int_0^t \sigma \int_s^t e^{-a(u-s)}du\, dW_s\right\}, \end{aligned} \tag{3.27}$$

where the boundedness of the integrand allows us to use the Fubini theorem[2] in the last equality.

We now attempt to write our martingale M so that bond-prices in our framework have the same volatility structure as in the above Hull-White model. First, in the

---

[2]For the generalisation of this result to stochastic integrals, see, for example, Protter [86].

framework of Sect. 2.3 we have that

$$D_{t,T} = \frac{D_{0,T}}{D_{0,t}} \frac{M_{t,T}}{M_{t,t}}, \tag{3.28}$$

and

$$N_t^{-1} = D_{0,t} M_{0,t} \exp\left\{ \int_0^t \sigma_{u,t} dW_{u,t} - \frac{1}{2} \Sigma_{t,t}^2 t \right\}, \tag{3.29}$$

where

$$\Sigma_{t,t}^2 := \frac{1}{t} \int_0^t \sigma_{u,t}^2 du. \tag{3.30}$$

Comparing (3.27) and (3.29) we can then set

$$\sigma_{t,t}^2 = \frac{\sigma^2}{a^2} \left(1 - e^{-at}\right)^2. \tag{3.31}$$

To identify the form of $\Sigma_{t,T}$ for $T \neq t$ in terms of the Hull-White parameters, we write $D_{t,T}$ in terms of the martingale $M_{t,T}$:

$$\log D_{t,T} = \log \frac{D_{0,T}}{D_{0,t}} + \log \frac{M_{0,T}}{M_{0,t}}$$
$$+ \left(\int_0^t \sigma_{u,T} dW_t - \frac{1}{2} \Sigma_{t,T}^2 t \right) - \left(\int_0^t \sigma_{u,t} dW_t - \frac{1}{2} \Sigma_{t,t}^2 t \right), \tag{3.32}$$

which means that

$$d\left[\log D_{t,T}\right] \doteq \left(\sigma_{t,T} - \sigma_{t,t}\right) dW_t, \tag{3.33}$$

where $\doteq$ signifies that the two sides differ by a term in dt only. On the other hand, in the Hull-White model, we can write

$$d\left[\log D_{t,T}\right] \doteq -\sigma B_{t,T} dW_t, \tag{3.34}$$

where

$$B_{t,T} := \frac{1}{a} \left(1 - e^{-a(T-t)}\right). \tag{3.35}$$

We therefore have

$$\sigma_{t,T} - \sigma_{t,t} = -\sigma B_{t,T} = -\frac{\sigma}{a} \left(1 - e^{-a(T-t)}\right) \tag{3.36}$$

with $\sigma_{t,t}$ as in (3.31).

Putting everything together, our framework can therefore be specialized to Hull-White volatility dynamics by choosing

$$\sigma_{t,T} = \left(\frac{\sigma}{a} e^{at} e^{-aT}\right) + \left(-\frac{\sigma}{a} e^{-at}\right). \tag{3.37}$$

Referring to the separable specification of volatility discussed in Sect. 3.1.1, the Hull-White model is achieved by having $n = 2$ and

$$\mathbf{f}_t = \text{Diag}\left(e^{at}, e^{-at}\right);$$

$$\mathbf{g}_T = \left(\frac{\sigma}{a}e^{-aT}, -\frac{\sigma}{a}\right)^T; \qquad (3.38)$$

$$\mathbf{R} = \frac{1}{\sqrt{2}}\begin{pmatrix} 1 & 1 \\ 1 & 1 \end{pmatrix}.$$

In effect, what happens here is that our general volatility function $\sigma_{t,T}$ is a linear combination of two separable forms. The degeneracy of the matrix $\mathbf{R}$ means that the components of the Brownian Motion $\mathbf{RW}$ are both equal to the single Brownian Motion $(W_1 + W_2)/\sqrt{2}$, consistent with there being only one source of noise as in the original Hull-White model.

3.2 Equity and FX Models

The vast literature on option-pricing models, stemming from the seminal Black-Scholes paper [15], consists of various attempts at dealing with the fact that stock prices and foreign exchange rates do not follow the simple dynamics of the Black-Scholes model.[3] In particular, financial asset price returns are not normally distributed (exhibiting skewness and fat tails), and volatilities vary with time and with the price level itself. In fact, it is market practice to quote option prices in terms of implied volatilities, that is, the value at which to evaluate the Black option-pricing formula that recovers the market option price.

At a general level, the devices for modelling these departures from asset price log-normality are, among others,

(i) Building a market model whereby stock and FX forward prices are modelled directly, employing Black-Scholes volatilities implied from observed option prices;

(ii) Making the volatility function depend on the level of the underlying, a class of models commonly referred to as *local volatility models*;

(iii) Allowing the volatility of the underlying process to be itself stochastic, thus introducing an additional source of noise. This class of models is referred to as *stochastic volatility models*;

(iv) Having a stock price process with a discontinuous component such as a Lévy process.

Our goal in this section is to specify equity and FX models within our modelling framework, and to show how they can be calibrated to market instruments. Recall

---

[3]For a survey of stylised facts about asset price returns, see, for example, Cont [29].

from Sect. 2.4 the expression

$$F_{t,T} = F_{0,T} \bar{F}_{t,T} = F_{0,T} Y_t \frac{\tilde{M}_{t,T}}{M_{t,T}} \qquad (3.39)$$

for the evolution of the FX forward price in terms of the martingales M and $\tilde{M}$ driving the reference and foreign rates of interest. Typically, models of the types enumerated above, designed to reproduce some desired feature of stock prices or foreign exchange rates, do away with complexity arising from discounting by assuming interest rates to be deterministic or constant. In terms of our framework this amounts to having $\tilde{M} \equiv M \equiv 1$ (whence $Y \equiv \bar{F}$), and therefore in what follows we will sketch the details of how the Black, local volatility and stochastic volatility models can be applied to the process Y. To see how this relates to the more familiar situation where the spot price is modelled directly, note that $Y = \bar{F}$ means $Y_t = F_{t,T}/F_{0,T}$. In particular, for $T = t$ we then have $S_t \equiv F_{t,t} = F_{0,t} Y_t$, so that

$$\frac{dS_t}{S_t} = \frac{\dot{F}_{0,t}}{F_{0,t}} dt + \frac{dY_t}{Y_t}; \qquad (3.40)$$

that is, imposing a volatility on Y amounts to doing the same on the spot price S.

Because we need to deal with all asset types (including interest-rate products) simultaneously, we also outline in Sect. 3.2.5 the form these models take when rates *are* stochastic, in which case the martingale terms in (3.39) contribute additional volatility to $\bar{F}$. The key idea will be to reduce the general case to the case of non-stochastic rates by taking an appropriate conditional expectation. We also look at the simpler approach, mentioned in (2.53), where we model forward rates (for equity and foreign exchange assets) that are independent of interest rates.

3.2.1 Black Model

The Black model extends the original Black-Scholes model, by writing local-martingale dynamics for asset forward prices, of the form

$$dY_t = Y_t g(t) dW_t, \qquad (3.41)$$

with W an $\mathbb{N}$-Brownian Motion and $g(t)$ a deterministic function in t.

In order that this specification of forward prices match observed option prices on the underlying asset, one simply compares the variance of Y in (3.41) to the observed Black-Scholes implied volatility for options of different maturities. That is, if maturities $T_1, \ldots, T_n$ are available, then g has to be consistent with

$$\left[\frac{1}{T_i} \int_0^{T_i} g_u^2 du \right]^{\frac{1}{2}} = \Sigma^{BS}(F_{0,T_i}, T_i), \quad i = 1, 2, \ldots n, \qquad (3.42)$$

where $\Sigma^{BS}(K, T)$ is the Black-Scholes implied volatility for an option of expiry T and strike K.

An alternative approach would be to calibrate to variance swap strikes, if a market for these is available. The constraint on g in this case is

$$\frac{1}{T_i} \int_0^{T_i} g_u^2 du = \mathbb{E}[\mathcal{V}_{T_i}], \tag{3.43}$$

where for each $t > 0$, the time-t realised variance $\mathcal{V}_t$ is defined to be

$$\mathcal{V}_t := \frac{1}{t} \int_0^t [\ln F_{u,t}]_u du, \tag{3.44}$$

in terms of the quadratic variation process,[4] $[\ln F]$, of $\ln F$.

Now, the realised variance can be expressed as an integral over call and put prices. Indeed[5] we find

$$\mathbb{E}[\mathcal{V}_T] = D_{0,T}^{-1} \frac{2}{T} \left(\int_0^{F_{0,T}} \frac{P_0(K, T)}{K^2} dK + \int_{F_{0,T}}^{+\infty} \frac{C_0(K, T)}{K^2} dK \right), \tag{3.45}$$

in terms of the time-zero prices of calls (resp. puts) $C_0(K, T)$ (resp. $P_{0,T}(K, T)$). In practice, there are subtleties in how to compute the fair variance swap strike, which we will not discuss here.

3.2.2 Local Volatility

The Black model of the previous section is calibrated to a different volatility for each at-the-money option maturity, but still makes the implicit assumption that options of different strikes exhibit the same implied volatility. In practice, implied volatilities observed in the market vary not just with option expiry but also with the strike of the option, an effect usually called a *smile* or *skew*. That is to say, market prices

---

[4]If M is a continuous local martingale, then there exists a unique increasing continuous process, $[M]$, called the quadratic variation process of M, such that $M^2 - [M]$ is a continuous local martingale. See Rogers & Williams [94] for the full story.

[5]Taylor's expansion series with integral remainder gives for any $C^2(\mathbb{R})$ function that

$$f(b) = f(a) + f'(a)(b - a) + \int_a^b f''(x)(b - x) dx$$

$$= f(a) + f'(a)(b - a) + \int_a^b f''(x)(b - x)^+ dx - \int_a^b f''(x)(x - b)^+ dx.$$

We apply this to $f(x) = \ln x$ with $a = F_{0,T}$ and $b = F_{T,T}$. Taking $\mathbb{N}$-expectations and noting that $\mathbb{E}[F_{T,T}] = F_{0,T}$ yields the result.

for options are inconsistent with assuming that the volatility of the underlying price process does not change with the price level.

One way of building this effect into the model is to modify the Black SDE (3.41) to

$$dY_t = g(t, Y_t)Y_t dW_t, \tag{3.46}$$

where the volatility function now depends also on Y. Models of the form (3.46) are referred to as *local-volatility* models. The departure of (3.46) from the Black model can be gauged by analysing the special case where the volatility term takes the form $g(t, x) = g(t) f(x)$. Expanding the function f around the value 1 and comparing to the implied volatility, we have (to second-order)

$$\int_1^m f(x)dx \approx \int_1^m \alpha + \beta(x - 1) + \frac{1}{2}\gamma(x - 1)^2 dx$$

$$\approx (m - 1)\frac{\Sigma^{BS}(m F_{0,T}, T)}{\Sigma^{BS}(F_{0,T}, T)}, \tag{3.47}$$

where $\alpha \equiv f(1)$, $\beta \equiv f'(1)$ and $\gamma \equiv f''(1)$, and where for each T, g relates to the at-the-money implied volatilities,

$$\int_0^T g^2(u)du = (\Sigma^{BS}(F_{0,T}, T))^2 T,$$

as in the Black model. Doing the integration, we see that we need to have

$$\frac{\Sigma^{BS}(m F_{0,T}, T)}{\Sigma^{BS}(F_{0,T}, T)} \approx \left(1 + \frac{1}{2}\beta(m - 1) + \frac{1}{6}\gamma(m - 1)^2\right), \tag{3.48}$$

which can be used in the Black formula to calculate approximate prices for options in this model, an observation that is important for calibration.

From (3.48), we interpret the coefficients β and γ in terms of the at-the-money skew and convexity, that is

$$\text{ATM Skew} = K\left.\frac{\partial \Sigma^{BS}(K, T)}{\partial K}\right|_{K=F_{0,T}} = \frac{1}{2}\beta\Sigma^{BS}(F_{0,T}, T) \tag{3.49}$$

$$\text{ATM Convexity} = K^2\left.\frac{\partial^2 \Sigma^{BS}(K, T)}{\partial K^2}\right|_{K=F_{0,T}} = \frac{1}{3}\gamma\Sigma^{BS}(F_{0,T}, T) \tag{3.50}$$

showing that if it is deemed suitable to have f quadratic, then (3.48) characterises f in terms of observed implied volatilities. We also read off Derman's approximate rule of thumb (see Derman [36]) that near $m = 1$, the local volatility f changes with m twice as fast (at a rate of β) as the implied volatility (which changes at a rate of $\frac{1}{2}\beta$).

The most well known example of a local volatility model is the Constant Elasticity of Variance (CEV) model, first studied by Cox & Ross [32] (see also

Schroder [97]). In our notation, the CEV model is obtained by choosing

$$g(t, x) = \sigma x^{\beta},$$ (3.51)

where $\sigma > 0$ is a positive real and $\beta \equiv f'(1)$ is a constant skewness parameter[6] as seen in (3.48).

For $\beta = 0$, the CEV model is obviously the Black Model in (3.41). For $\beta < 0$ (resp. $\beta > 0$), the volatility decreases (resp. increases) with x. This results in the distribution of Y_t being skewed to the left (resp. to the right), as shown in Fig. 3.1.

Fig. 3.1 Distribution of $F_{T,T} = S_T$ with $F_{0,T} = 100$, $\sigma = 20\%$ and $T = 1$

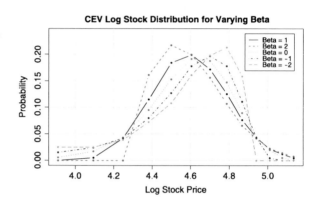

Taken at face value, local volatility models impose on asset prices a volatility term that is a deterministic function of the asset price level, an assumption that might be questionable in practice. However, Dupire [41] showed the existence of a *diffusion process* consistent with the observed local volatility surface, and the diffusion coefficient for this process is a local volatility function. We review the derivation of this result, following closely the analysis of Dupire's idea in Gatheral [48].

Consider the time-zero price, $C(F_{0,T}, mF_{0,T}, T)$ say, of an option of expiry T and strike $mF_{0,T}$, where $F_{0,T}$ is the time-zero price of the T-forward and $m > 0$ is the *strike moneyness*. If we write $p_t(\cdot, \cdot)$ for the risk-neutral transition density of Y, the call price has the representation

$$C(F_{0,T}, mF_{0,T}, T) = \int_m^{\infty} D_{0,T}(yF_{0,T} - mF_{0,T})p_T(1, y)dy.$$ (3.52)

The Kolmogorov equations for the transition density $p_T(\cdot, \cdot)$ read

$$\dot{p}_t(x, y) = \mathscr{G}_x p_t(x, y) = \mathscr{G}_y^* p_t(x, y),$$ (3.53)

---

[6]The parameter β is also related to the elasticity of the variance in this model, which is $x(f^2)'(x)/f^2(x) = 2\beta$.

where $\mathcal{G}$, $\mathcal{G}^*$ are, respectively, the infinitesimal generator[7] of Y and its adjoint, and where $\dot{p}_t$ represents differentiation in t. Differentiating (3.52) in T and using the first part of (3.53), we obtain the Black-Scholes differential equation. Because we are interested in a differential equation involving derivatives in the strike and not in the initial forward price $F_{0,T}$, we use the second part of (3.53) to get

$$\dot{C}(y,t) = \left(\frac{\dot{D}_{0,t}}{D_{0,t}} + \frac{\dot{F}_{0,t}}{F_{0,t}}\right) C(y,t) - \frac{\dot{F}_{0,t}}{F_{0,t}} y C'(y,t)$$

$$+ \frac{1}{2} g^2(t,y) y^2 C''(y,t), \tag{3.54}$$

where $C(y,t)$ is an abbreviation for $C(F_{0,T}, y F_{0,T}, t)$, C' and C'' are derivatives of C in its argument y, and $\dot{C}$ is the derivative in t. Inverting (3.54) gives us an expression for the local volatility function g in terms of option moneyness and maturity, that is

$$g^2(t,y) = \frac{\dot{C}(y,t) - \left(\frac{\dot{D}_{0,t}}{D_{0,t}} + \frac{\dot{F}_{0,t}}{F_{0,t}}\right) C(y,t) + \frac{\dot{F}_{0,t}}{F_{0,t}} y C'(y,t)}{\frac{1}{2} y^2 C''(y,t)}. \tag{3.55}$$

Although C and the derivatives C' and C'' are related to market call prices, the final goal will be reached only when we express g in terms of what is directly observed in the market, namely the *implied volatility surface*. Gatheral [48] shows the required expression to be

$$g^2(t,y) = \dot{w} \left/ \left[1 - \frac{\varphi}{w}\frac{\partial w}{\partial \varphi} + \frac{1}{4}\left(-\frac{1}{4} - \frac{1}{w} + \frac{\varphi^2}{w^2}\right)\left(\frac{\partial w}{\partial \varphi}\right)^2 + \frac{1}{2}\frac{\partial^2 w}{\partial \varphi^2}\right], \right. \tag{3.56}$$

where

$$w(F_{0,T}, y, T) = T \Sigma^{BS}(F_{0,T}, y F_{0,T}, T), \tag{3.57}$$

is the Black-Scholes implied total variance,

$$\dot{w} \equiv \partial w / \partial T, \tag{3.58}$$

and where

$$\varphi \equiv \ln y \tag{3.59}$$

is the log-moneyness of the option strike.

In practice, to get around difficulties arising from the volatility surface not being smooth or granular enough, one would parametrize the total variance surface w in

---

[7] If X is a diffusion, then for smooth f the process $Y_t = f(X_t) - \int_0^t (\mathcal{G} f)(X_s) ds$ is a martingale. $\mathcal{G}^*$ is then defined by having $\int g(x)(\mathcal{G} f)(x) dx = \int (\mathcal{G}^* g)(x) f(x) dx$ for smooth f and g. From this characterisation, we see that Y is a martingale, supermartingale or submartingale according as f is harmonic ($\mathcal{G} f = 0$), superharmonic ($\mathcal{G} f \leq 0$), or subharmonic ($\mathcal{G} f \geq 0$) for $\mathcal{G}$.

terms of powers of moneyness $K/F_{0,T}$, and use derivatives of this parametrized form in computing (3.56). Figure 3.2 shows the implied volatility as a function of strike for chosen values of skew and convexity. Similarly, Fig. 3.3 shows the effect of skew on the resulting distribution for the log-price of the stock.

3.2.3 Stochastic Volatility

The class of *stochastic volatility* models allows volatility of an asset price and the asset price itself to be altogether different processes, by writing dynamics of the form

$$dY_t = Y_t \sqrt{v_t} dW_t \tag{3.60}$$

$$dv_t = \alpha(Y_t, v_t, t)dt + \beta(Y_t, v_t, t)dZ_t \tag{3.61}$$

with W and Z being $\mathbb{N}$-Brownian Motions having instantaneous correlation ρ.

Now from Dupire's result we know that prices of options (equivalently, the implied volatility surface) of some given expiry T are consistent with a distribution of asset prices at T arising from some local-volatility model. In fact, it is a consequence of a more general result of Gyöngi [54] that if the SDE for Y in (3.60) admits a unique solution, then the SDE

$$dX_t = X_t b(t, X_t)dW_t, \qquad X_0 = Y_0, \tag{3.62}$$

has a weak solution[8] having the same law as Y. Moreover, the diffusion term b has the representation

$$b^2(t, y) = \mathbb{E}[v_t \mid Y_t = y], \tag{3.63}$$

thus exposing the link between the stochastic volatility process v and the Dupire local volatility function b. For more details on diffusion and stochastic volatility models we refer the reader to Andersen & Piterbarg [3], Cox [31], Dupire [41], Derman & Kani [37], and Hagan et al. [55].

A well-known example of a stochastic volatility model is the Heston model (see Heston [61]), in which the asset price variance is a diffusion of CIR type, that is

$$dv_t = \kappa(\theta - v_t)dt + \sigma\sqrt{v_t}dZ_t. \tag{3.64}$$

This diffusion is mean-reverting to a level $\theta > 0$, with $\kappa > 0$ controlling the speed of reversion. The real parameter $\sigma > 0$ is commonly known as the *volatility of volatility*. Correlation between the Brownian Motions Z (driving volatility) and W (driving the asset price Y) controls the implied volatility skew, with negative skew resulting

---

[8]For the concept of a weak solution to an SDE, see any standard text on the subject, such as Rogers & Williams [94].

from negative correlation (much like negative β in the CEV model). Euler and Milstein schemes are not the best choice for discretizing and simulating the pair of SDEs (3.60)–(3.61); Andersen [1] presents an efficient method that makes use of moment-matching techniques.

For the Heston model (part of a bigger class of generally tractable models—see Duffie et al. [40]), the characteristic function of the asset price distributions is known in closed form. By this, and as shown by Carr & Madan [24], the characteristic function of the option price can itself be written down in closed form, allowing option prices to be recovered by numerical inversion. In general, if $s_t = \ln(Y_t)$ and

$$\phi_T(u) = \mathbb{E}[\exp(ius_T)] \quad (i^2 \equiv -1), \tag{3.65}$$

is the characteristic function of s_T, then the modified call prices

$$c(m, T; \zeta) \equiv F_{0,T}^{-1} e^{\zeta m} C(F_{0,T}, e^m F_{0,T}, T) \tag{3.66}$$

with strike K, log strike moneyness $m = \ln(K/F_{0,T})$, maturity T and real $\zeta > 0$, have Fourier transform

$$\psi_T(v) \equiv \int_{-\infty}^{\infty} e^{ivx} c(x, T; \zeta) dx = \frac{D_{0,T} \phi_T(v - (\zeta + 1)i)}{\zeta^2 + \zeta - v^2 + i(2\zeta + 1)v}, \tag{3.67}$$

which shows why positivity of ζ is required to avoid having ψ_T singular at the origin.

Fourier techniques such as FFT can be used to calibrate the model using observed call prices, by numerically inverting the option price transform (3.67). In the particular case of the Heston model[9] the transform (3.65) is log-affine in the starting point V_0 of the variance process V, namely

$$\phi_T(u) = \exp\{C(T, u) + D(T, u)V_0\}, \tag{3.68}$$

where

$$C(T, u) = \frac{\kappa\theta}{\sigma^2} \left[(\kappa - \rho\sigma ui + d(u)) T - 2\ln\left(\frac{c(u)e^{d(u)T} - 1}{c(u) - 1}\right) \right],$$

$$D(T, u) = \frac{\kappa - \rho\sigma ui + d(u)}{\sigma^2} \left(\frac{e^{d(u)T} - 1}{c(u)e^{d(u)T} - 1}\right), \tag{3.69}$$

with

$$c(u) = \frac{\kappa - \rho\sigma ui + d(u)}{\kappa - \rho\sigma ui - d(u)}, \qquad d(u) = \sqrt{(\rho\sigma ui - \kappa)^2 + (iu + u^2)\sigma^2}. \tag{3.70}$$

Note that in (3.68), we have $\phi_T(-i) = \mathbb{E}[Y_T] = 1$, because $C(T, -i) = D(T, -i) = 0$; this is of course as expected from the martingale property of Y.

---

[9] See, for example, Kahl & Jaeckel [67].

3.2.4 Jump Models

Processes with jumps are an attractive tool for modelling asset prices, with Merton [81] being the first to explore option pricing in models where the stock price has a discontinuous component. Geman et al. [49] and Carr et al. [26] present a case for using pure jump processes (that is, with no diffusion component) to model asset price returns. Linked to this idea is the use of non-decreasing jump processes[10] to time-change continuous diffusions, thus providing a powerful device for matching the implied volatility skew (see also Carr & Wu [25], Eberlein et al. [42], Kou [71], and Mendoza et al. [79]).

The basic class of jump processes with which anything tractable can be done is Lévy processes, because the infinite divisibility property gives the characteristic function a form that enables calculation of various Laplace transforms and therefore opens the way to inversion of option prices as for the Heston model. For more information on Lévy processes, we refer the reader to Barndorf-Nielsen et al. [9], Bertoin [13], Boyarchenko & Levendorskii [16], and Rogers & Williams [94]. For work on models combining stochastic volatility with Lévy processes see, for example, Carr et al. [27]. There is a vast literature on the use of jump processes in financial modelling, of which Cont & Tankov [30] is a good survey. Schoutens & Symens [96] studies the pricing of options by simulation in jump models.

3.2.5 Extension to Stochastic Interest Rates

The previous sections have reviewed standard specifications for the Black, local volatility and stochastic volatility models. Because the primary goal of these models is to explain or fit observed volatilities implied from option prices, the complexities brought about by stochastic interest rates are avoided by assuming rates and bond prices to be deterministic. The assumption of deterministic rates is a severe constraint if models are to be used for credit exposure computation, but what we have done above has not been in vain. In fact this section will describe how, by introducing an extra conditioning step, the models for deterministic interest rates can be re-used to take into account stochastic interest rates.

In our framework, where we have (see (3.39))

$$F_{t,T} = F_{0,T} \bar{F}_{t,T} = F_{0,T} Y_t \frac{\tilde{M}_{t,T}}{M_{t,T}}, \qquad (3.71)$$

deterministic rates are equivalent to the martingales M and $\tilde{M}$ being identically equal to one, and the asset pricing models as we have described them then apply to the process $Y_t = F_{t,T}/F_{0,T}$, where Y_T then has the interpretation as the moneyness of the time-T spot price relative to the time-zero forward price.

---

[10]Real-valued increasing Lévy processes are termed *subordinators*.

We now review how such models can be extended to the case when the rates martingales are not deterministic, similar to what is done by Andreasen in [5], and assuming throughout that M and $\tilde{M}$ follow log-normal dynamics. The key idea is to write the process Y as the product $Y^{\vdash} Y^{\perp}$ of two independent martingales. Conditioning on the value of $Y^{\vdash}$, an option on $Y^{\perp}$ can be priced in the stochastic or local volatility models, with the final option price then being an integral over the law of $Y^{\vdash}$ of the conditional option prices.

To see this, in (3.71) we start by writing

$$\bar{F}_{t,T} = Y_t \frac{\tilde{M}_{t,T}}{M_{t,T}} =: \left[Y_t^{\vdash} \frac{\tilde{M}_{t,T}}{M_{t,T}} \right] Y_t^{\perp} \equiv \bar{F}_{t,T}^{\vdash} Y_t^{\perp}, \tag{3.72}$$

with $Y_t^{\vdash}$ *independent* of $Y_t^{\perp}$ (but not, of course, independent of $(M, \tilde{M})$[11]). Intuitively, as the notation is intended to help convey, $Y^{\perp}$ relates to the stochastic component in asset forward prices that is 'orthogonal' to rates. Of course, for deterministic rates, $Y^{\perp}$ coincides with Y and $Y^{\vdash}$ is identically equal to one.

The advantage of this decomposition is that it allows to calibrate the model in two distinct steps, separating the asset price and interest rate components, as we see now. Consider a call option of maturity T and strike $K = m F_{0,T}$, which has price

$$C(F_{0,T}, m F_{0,T}, T) = \mathbb{E}\left[N_T^{-1} \left(F_{T,T} - m F_{0,T} \right)^+ \right]. \tag{3.73}$$

Conveniently switching to the T-forward measure, we have

$$C(F_{0,T}, m F_{0,T}, T) = D_{0,T} \mathbb{E}^{\mathbb{N}^T}\left[\left(\bar{F}_{T,T} F_{0,T} - m F_{0,T} \right)^+ \right]$$

$$= D_{0,T} F_{0,T} \mathbb{E}^{\mathbb{N}^T}\left[\left(\bar{F}_{T,T} - m \right)^+ \right]$$

$$= D_{0,T} F_{0,T} \mathbb{E}^{\mathbb{N}^T}\left[\bar{F}_{T,T}^{\vdash} \left(Y_T^{\perp} - m / \bar{F}_{T,T}^{\vdash} \right)^+ \right]. \tag{3.74}$$

The crucial next step is now to invoke the independence of $Y^{\vdash}$ and $Y^{\perp}$ to develop the call price as

$$C(F_{0,T}, m F_{0,T}, T)$$

$$= F_{0,T} \mathbb{E}^{\mathbb{N}^T}\left[\bar{F}_{T,T}^{\vdash} D_{0,T} \mathbb{E}^{\mathbb{N}^T}\left[\left(Y_T^{\perp} - m / y^{\vdash} \right)^+ \middle| y^{\vdash} = \bar{F}_{T,T}^{\vdash} \right] \right]$$

$$= F_{0,T} \mathbb{E}^{\mathbb{N}^T}\left[\bar{F}_{T,T}^{\vdash} C^{\perp} \left(1, m / \bar{F}_{T,T}^{\vdash}, T \right) \right], \tag{3.75}$$

where $C^{\perp}(y^{\perp}, k, T)$ is the price of an option on $Y^{\perp}$ (whose initial value is $y^{\perp}$) struck at k, of expiry T.

---

[11] We emphasise again that, for tractability reasons, we will assume $Y_t^{\vdash}$ follows a log-normal process.

The usefulness of decomposing Y is now clear: to compute the price of an option on F, we first compute prices of an option on $Y^\perp$, using some preferred asset-pricing model and unencumbered by rates stochasticity, and *then* incorporate the effect of rates by integrating over the law of $\bar{F}^\perp$.

If we take log-normal Black models for each of $Y^\perp$, M and $\tilde{M}$, then $\bar{F}^\perp$ will be also log-normal (and a $\mathbb{N}^T$-martingale). We then have the equality in law

$$\bar{F}^\perp_{T,T} \sim \exp\left(Z\sqrt{T}\,\Sigma^\perp_T - \frac{1}{2}(\Sigma^\perp_T)^2 T\right) \equiv \mathscr{E}(Z, T, \Sigma^\perp_T) \quad (Z \sim N(0,1)), \quad (3.76)$$

where the total variance $T(\Sigma^\perp_T)^2$ is a function of the volatilities and covariance matrix of $Y^\perp$, M and $\tilde{M}$. Thus, if $\mathbf{g}_t = (g^\perp_t, \tilde{g}_t, g_t)$ is the vector of Black volatilites of $(Y^\perp_t, \tilde{M}, M)$, then we have

$$T(\Sigma^\perp_T)^2 = \int_0^T \mathbf{g}_u \cdot \mathbf{R}\mathbf{g}_u\, du, \quad (3.77)$$

where, with obvious notation, the matrix

$$\mathbf{R} = \begin{pmatrix} 1 & -\rho_{Y,\tilde{M}} & \rho_{Y,M} \\ -\rho_{Y,\tilde{M}} & 1 & \rho_{M,\tilde{M}} \\ \rho_{Y,M} & \rho_{M,\tilde{M}} & 1 \end{pmatrix} \quad (3.78)$$

has as entries the pair-wise correlations between Brownian Motions driving $Y^\perp$, M and $\tilde{M}$.

With this setup, we can now write the call price fairly explicitly as a Gaussian integral, for we have from (3.75) that

$$C(F_{0,T}, mF_{0,T}, T)$$

$$= D_{0,T}F_{0,T}\int_{-\infty}^\infty \mathscr{E}(z, T, \Sigma^\perp_T)C^\perp\left(1, m/\mathscr{E}(z, T, \Sigma^\perp_T), T\right)\phi(z)dz$$

$$= D_{0,T}F_{0,T}\int_{-\infty}^\infty C^\perp\left(1, m/\mathscr{E}(z + \sqrt{T}\,\Sigma^\perp, T, \Sigma^\perp_T), T\right)\phi(z)dz, \quad (3.79)$$

where the last equality follows immediately once we interpret the 'Doléans' exponential[12] $\mathscr{E}$ defined in (3.76) as a change of measure.

The integral in (3.79) readily lends itself to numerical methods (such as gaussian quadrature) as long as the call price is available. In the case of stochastic volatility models such as the Heston model, call prices are known only up to a Fourier transform, so FFT methods will be needed to compute the call prices by inversion, as discussed in Sect. 3.2.3.

---

[12]For a martingale M with $M_0 = 0$, the Doléans exponential of M is the exponential martingale process $\mathscr{E}_t(M) = \exp(M_t - \frac{1}{2}[M]_t)$, where $[M]$ is the quadratic variation process of M. Our use of the name is because the law of (3.76) is the law of $\mathscr{E}_T(\Sigma^\perp_T B)$ for some Brownian motion B.

There is nothing to stop us from decomposing the process $Y^\perp$ itself as the product of independent processes, as long as the characteristic functions of the time-t laws of those processes (and hence the characteristic functions of call option prices) are known in closed form. This allows the pricing of options in models where not only are rates stochastic, but also where different features of the volatility surface can be accommodated by employing independent local volatility, stochastic volatility and even jump components, in the spirit of Andreasen's [4, 5].

3.2.6 A Simpler Approach: Independent Interest Rates

It is worth contrasting the decomposition of Y in (3.72), whereby

$$Y_t^\perp = \bar{F}_{t,T} \left[Y_t^\vdash \frac{\tilde{M}_{t,T}}{M_{t,T}} \right]^{-1} , \quad Y^\vdash, Y^\perp \text{ independent,} \tag{3.80}$$

to the approximative one of (2.53), where we had simply

$$Y_t \approx \hat{F}_{t,T} \left[\frac{\tilde{M}_{t,T}}{M_{t,T}} \right]^{-1} , \quad \hat{F}, (\tilde{M}, M) \text{ independent,} \tag{3.81}$$

with $\hat{F}$ and $\bar{F}$ having the same *marginal* laws. Expression (2.54) showed how the inaccuracy brought about by (3.81) can be mitigated somewhat, and in Chap. 6 we will analyse by means of concrete examples the impact this inaccuracy has on pricing. Benhamou [11] analyses the bias introduced by neglecting the stochasticity of interest rates when deriving the Dupire formula.

3.2.7 Different Models for Different Markets

Certain models may be better suited to equity markets than to FX markets. For example, equity markets usually have negative skews, whereas FX markets tend to have smiles. In other words, while the market-implied volatilities in equity markets are impacted by skew, those in FX markets are impacted by kurtosis. One possible economic reason for this is that traditional investors in equities can only be long, whereas one can always go short in foreign exchange markets by switching one's holdings from one currency to another.

Figure 3.2 shows typical implied volatilities for equities and foreign exchange markets. Figure 3.3 shows how the shape of the implied volatility surface impacts the distribution of log-FX rates.

As described in Chap. 2, inflation markets can be seen as an extension of foreign exchange markets, with the exception that inflation markets are less complete. Indeed, while options on FX rates are commonplace, options on inflation indices

Fig. 3.2 Typical implied volatilities for Equity Markets (at-the-money skew = −10%, at-the-money convexity = 0%) and Foreign Exchange Markets (at-the-money skew = 0%, at-the-money convexity = 30%, see (3.49))

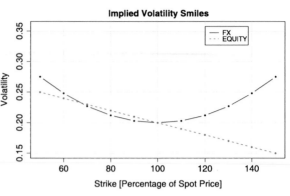

Fig. 3.3 Implied distributions for Foreign Exchange Markets (above: at-the-money skew = 0%, at-the-money convexity = 30%) and Equity Markets (below: at-the-money skew = −10%, at-the-money convexity = 0%) compared to the standard normal distribution, see (3.49)

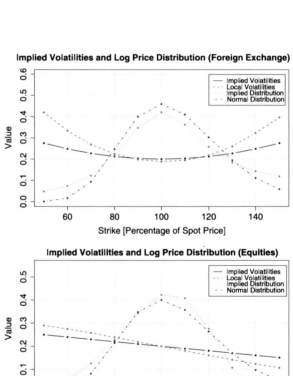

are not. Therefore using complex dynamics for inflation indices and real rates may be unnecessary, or even counterproductive since there are not enough liquid instruments available to calibrate the model.

3.3 Credit Models

Credit Derivatives are products whose payoff is related to credit quantities such as credit spreads, credit default losses, or rating migrations. For credit derivatives relating to a portfolio of more than one entity, an essential modelling element is the inter-dependence between the spread and default times of the individual entities. Indeed, computing prices and simulating price distributions for credit products is challenging as (i) the choice is not clear as to what the best model is for simulating credit spreads and default times, and (ii) there are several ways of introducing default dependence between different credit entities (see for example Duffie & Singleton [39], Lando [72], or Schönbucher [95]).

This section describes a possible model for computing the credit exposure posed by credit-related products. The starting point, as described in Sect. 2.7, is to use CDS term structure to compute values at time zero of the default probabilities for each single reference name. Once this is done, stochastic default probabilities for each reference name can be simulated following some chosen dynamics for the martingales $\bar{M}$ appearing in Sect. 2.7.2; this is what we do in Sect. 3.3.1 below. We also describe how the volatility parameters in the model can be calibrated to market quotes for options on CDSs. For the dependence between different reference entities, we have chosen to work with a Gaussian dependence structure, as hinted at in (2.72). We detail in this section how we propose to calibrate this dependence structure to quotes on CDO tranches.

At this point it is worth highlighting the difficulties one encounters in simulating price distributions of credit portfolios. When faced with the task of *pricing* correlation-dependent products such as, say, CDO tranches, it is standard practice to match market prices for different tranches by tuning model parameters (usually the correlation input) individually for each tranche.[13] Such an approach is useless, however, in a simulation model attempting to produce correct default loss distributions for different tranches *simultaneously*. Consequently, the correlation structure in our model has no meaning beyond that given to it in the modelling expressions (2.72).

In what follows, we consider a portfolio of n defaultable entities, and let A_i, R_i and $\tau^{(i)}$ respectively define the nominal amount, the recovery fraction upon default, and the time of default of name i. Recall from (2.74) the portfolio loss process $L \equiv (L_t)_{t \geq 0}$,

$$L_t = \left[\sum_{i=1}^{n} (1 - R_i) A_i \mathbb{1}_{\tau^{(i)} \leq t} \right] / \sum_{i=1}^{n} A_i. \tag{3.82}$$

The quantity

$$q_{s,t}^{(i)} \equiv (1 - p^{(i)}(s, t)) = \mathbb{N}\left(\tau^{(i)} \geq t \mid \mathscr{F}_s, \tau^{(i)} > s \right), \tag{3.83}$$

---

[13] Often this parameter is called *base correlation* by practitioners.

will denote the probability, based on time-s information, that name i survives beyond time t, having survived until time $s < t$.

3.3.1 Simulation of Single-Name Default Probabilities and Default Times

Here we specify concrete dynamics for the martingale process $\bar{M}$ in (2.69). We choose a volatility term in the SDE for $\bar{M}$ that has a separable dependence on time and maturity, and which leads to normally-distributed survival probabilities.[14] More specifically, we set

$$d\bar{M}_{t,T} = f_t g_T dW_t, \qquad \bar{M}_{0,T} = q_{0,T}, \tag{3.84}$$

or, in integral form,

$$\bar{M}_{t,T} = q_{0,T} + g_T X_t \sim N(q_{0,T}, g_T^2 F_t^2 t), \tag{3.85}$$

with

$$X_t = \int_0^t f_u dW_u, \tag{3.86}$$

and with

$$F_t^2 = t^{-1} \int_0^t f_u^2 du \tag{3.87}$$

being the normalised variance of the time-changed process X. Note that the separability of the volatility coefficient of $\bar{M}_{t,T}$ is essential to allow fast access to any martingale value and therefore to any needed quantity (for instance, a par CDS spread).

We now look at how to parametrize the functions f and g and calibrate them to prices of options on CDSs. Consider an option, expiring at time $T \equiv T_0$, to enter into a CDS contract that pays protection on a chosen reference name in return for a fixed strike coupon K at coupon payment dates $T_1, T_2, \ldots, T_n$. Assuming a fixed CDS recovery rate of $R \equiv 1 - \bar{R}$, the no-arbitrage time-zero price of such an option is

$$\text{CDSS}(0, K, T, T_n) = \mathbb{E}\left[\left(\sum_{i=1}^n \bar{M}_{T,T} \frac{D_{T,T_i}}{N_T} \left\{\bar{R}\left(q_{T,T_{i-1}} - q_{T,T_i}\right) - \alpha_i K q_{T,T_i}\right\}\right)^+\right], \tag{3.88}$$

---

[14]We will see later in this chapter how to put bounds on the proportion of simulated probabilities not in [0, 1].

with $\alpha_i \equiv (T_i - T_{i-1})$. Expressing now the bond D_{T,T_i} in terms of the martingale M_{T,T_i} and the survival probability $q_{T,T_{i-1}}$ in terms of $M_{T,T_{i-1}}$, we get

$$\text{CDSS}(0, K, T, T_n) = \mathbb{E}\left[\left(\sum_{i=1}^{n} \frac{D_{T,T_i}}{N_T}\{\bar{R}\bar{M}_{T,T_{i-1}} - (\bar{R} + \alpha_i K)\bar{M}_{T,T_i}\}\right)^+\right].$$
(3.89)

This formulation allows one to price semi-analytically any CDS option, as long as the expectation in (3.89) can be computed in closed form. This is the case for the normal model in (3.84). Indeed, in (3.89), because $\bar{M}$ is normally distributed, so is the difference $(\bar{M}_{T,T_{i-1}} - \bar{M}_{T,T_i})$. Thus (neglecting rates stochasticity),

$$\text{CDSS}(0, K, T, T_n) = \mathbb{E}\left[\left(\sum_{i=1}^{n} D_{0,T_i}\left(\gamma_i(q_{i-1}, q_i) + \gamma_i(g_{i-1}, g_i)F_T\sqrt{T}Z\right)\right)^+\right],$$
(3.90)

where we have abbreviated $g_i \equiv g_{T_i}$, $q_i \equiv q_{0,T_i}$, $Z \sim N(0, 1)$, and where

$$\gamma_i(x, y) \equiv \bar{R}x - (\bar{R} + \alpha_i K)y.$$
(3.91)

The sum appearing in (3.90) is just a normal random variable.[15] We can write (3.90) as

$$\text{CDSS}(0, K, T, T_n) = \mu\Phi(\mu/\eta) + \eta\phi(\mu/\eta),$$
(3.92)

where μ and η are the mean and standard deviation, respectively, of the sum in (3.90), i.e.

$$\mu \equiv \sum_{i=1}^{n} D_{0,T_i}\gamma_i(q_{i-1}, q_i)$$

$$\eta \equiv \sum_{i=1}^{n} D_{0,T_i}|\gamma_i(g_{i-1}, g_i)|F_T\sqrt{T}.$$
(3.93)

In what we have above, $\bar{M}_{t,T}$ is a normal variate centered around $q_{0,T}$. It is desirable to have the martingale $\bar{M}$ stay within the interval $[0, 1]$ with high probability. This condition puts constraints on the forms one can choose for the volatility functions f and g. In more detail, if we want

$$0 \leq \bar{M}_{t,T} \leq 1$$
(3.94)

---

[15]Note that given a random variable $Z \sim N(0, 1)$, $a \in \mathbb{R}$, and $b > 0$, we can write

$$\mathbb{E}[a \pm bZ]^+ = a\Phi(a/b) + b\phi(a/b),$$

where ϕ is standard normal density and Φ is its primitive.

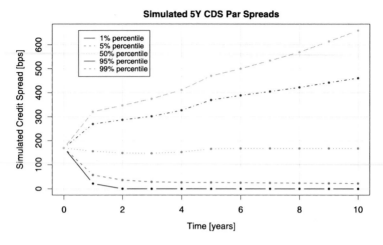

Fig. 3.4 Quantiles of 5Y CDS spreads resulting from simulating survival probabilities according to the separable model in (3.84). Model parameters for the volatility functions f and g were $(\kappa, \kappa_g, \sigma) = (0.4, 0, 1)$

to be true with probability p, then f and g will need to satisfy

$$q_{0,T} + q g_T F_t \sqrt{t} \le 1$$
$$0 \le q_{0,T} - q g_T F_t \sqrt{t}, \tag{3.95}$$

where $q = \Phi^{-1}(\frac{1}{2}(p+1))$. Both these conditions can be summarised as

$$g_T F_t \sqrt{t} \le \frac{\min(q_{0,T}, \bar{q}_{0,T})}{q}, \tag{3.96}$$

$\bar{q}_{0,T} = 1 - q_{0,T}$. Since the right hand side above involves only T, it is preferable to work on the assumption that F is known and then to ensure that

$$g_T \le \frac{\min(q_{0,T}, \bar{q}_{0,T})}{q \max_{t>0}(F_t \sqrt{t})}; \tag{3.97}$$

one satisfactory choice is

$$g_T = \sigma \exp(-\kappa_g T) \frac{\min(q_{0,T}, \bar{q}_{0,T})}{q \max_{t>0}(F_t \sqrt{t})} \tag{3.98}$$

with $\kappa_g \ge 0$ and $\sigma \le 1$.

From (3.84), choosing, then, say,

$$f_t = \exp(-\kappa t), \quad \kappa \ge 0, \tag{3.99}$$

whence

$$t F_t^2 = (1 - e^{-2\kappa t})/(2\kappa) \quad (\le (2\kappa)^{-1}), \tag{3.100}$$

we have finally the specific form of g as

$$g_T = q^{-1}\sigma\sqrt{2\kappa}\exp\left(-\kappa_g T\right)(\min(q_{0,T}, \bar{q}_{0,T})). \qquad (3.101)$$

We can now use this parametrization to compute options on CDS (as described above) and we can derive Black implied volatilities. In this model Black-implied volatilities of CDS option prices can be seen to

(i) decrease with option expiry;
(ii) increase as a function of maturity of the underlying CDS, then decrease again;
(iii) increase with the level of par CDS spread.

Figure (3.5) displays implied volatilities for options of different expiries, exercising into CDSs of varying maturities.

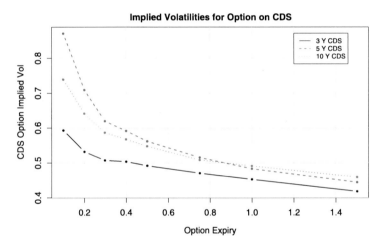

Fig. 3.5 Black-implied volatilities for options of various expiries giving the holder the right to enter into CDSs of different maturities. Volatility generally decreases with option expiry (x-axis). It increases to a peak (for the CDS of $5Y$ maturity) and then decreases again with maturity of the underlying CDS. Model parameters used were $(\kappa, \kappa_g, \sigma) = (0.4, 0, 1)$

3.3.2 Inter-Name Default Dependence

In the spirit of the Gaussian dependence model prescribed in (2.72), and in the context of the normal model for M, we now write down expressions for the joint law of default times of different entities; these are of course a function of the correlation parameters ρ and η in (2.72). In turn, these joint laws will be used to compute semi-analytically the prices of the CDO tranches that are chosen as calibration instruments, allowing calibration of the correlation structure to market information.

Consider the martingale process $\bar{M}^{(i)}$ driving the stochastic default probabilities for reference name i, namely

$$d\bar{M}_{t,T}^{(i)} = f_t^{(i)} g_T^{(i)} dW_t^{(i)}, \quad \bar{M}_{0,T}^{(i)} = q_{0,T}^{(i)}. \tag{3.102}$$

We will force the $M^{(i)}$ to depend on each other by decomposing the Brownian Motion $W^{(i)}$ as

$$dW_t^{(i)} = \boldsymbol{\eta}^{(i)} \cdot d\mathbf{Z}_t + \bar{\eta}^{(i)} dZ_t^{(i)}, \tag{3.103}$$

as in (2.72). $\mathbf{Z}_t$ is a d-dimensional Brownian Motion and $Z_t^{(i)}$ is a univariate independent Brownian Motion; $\bar{\eta}^{(i)} = \sqrt{1 - \boldsymbol{\eta}^{(i)} \cdot \boldsymbol{\eta}^{(i)}}$. This results in

$$\begin{aligned}
X_t^{(i)} &= \int_0^t f_u dW_u^{(i)} \\
&= \int_0^t f_u \left[\boldsymbol{\eta}^{(i)} \cdot d\mathbf{Z}_t + \bar{\eta}^{(i)} dZ_t^{(i)} \right] \\
&\sim F_t^{(i)} \sqrt{t} \left[\boldsymbol{\eta}^{(i)} \cdot \mathbf{Z} + \bar{\eta}^{(i)} Z^{(i)} \right],
\end{aligned} \tag{3.104}$$

where $t F_t^2$ is the variance of X and where $\mathbf{Z}$, $Z^{(i)}$ now denote random variables with same laws as $\mathbf{Z}_t / \sqrt{t}$, $Z_t^{(i)} / \sqrt{t}$, respectively.

Similarly, we force the uniform variables[16] $U^{(i)}$ to depend on each other by setting

$$U^{(i)} = 1 - \Phi \left(\boldsymbol{\rho}^{(i)} \cdot \mathbf{M} + \bar{\rho}^{(i)} M^{(i)} \right), \tag{3.105}$$

again as in (2.72).

With this setup, we can now write down the joint law of default times of different entities. Indeed, the cumulative distribution function of the default time of entity i is

$$\begin{aligned}
\mathbb{N}\left(\tau^{(i)} < t \right) &= \mathbb{N}\left(\bar{M}_{t,t}^{(i)} < U^{(i)} \right) \\
&= \mathbb{N}\left(q_{0,t}^{(i)} + g_t^{(i)} X_t^{(i)} \le U^{(i)} \right) \\
&= \mathbb{N}\left(1 - U^{(i)} \le (1 - q_{0,t}^{(i)}) - g_t^{(i)} X_t^{(i)} \right) \\
&= \mathbb{N}\left(\Phi \left(\boldsymbol{\rho}^{(i)} \cdot \mathbf{M} + \bar{\rho}^{(i)} M^{(i)} \right) \le \bar{p}_{0,t}^{(i)} - g_t^{(i)} X_t^{(i)} \right) \quad (\bar{p}_{0,t}^{(i)} \equiv 1 - q_{0,t}^{(i)}) \\
&= \mathbb{N}\Big(\Phi \left(\boldsymbol{\rho}^{(i)} \cdot \mathbf{M} + \bar{\rho}^{(i)} M^{(i)} \right) \\
&\quad \le \bar{p}_{0,t}^{(i)} - g_t^{(i)} \sqrt{t} F_t^{(i)} \left(\boldsymbol{\eta}^{(i)} \cdot \mathbf{Z} + \bar{\eta}^{(i)} Z^{(i)} \right) \Big)
\end{aligned}$$

---

[16] . . . from which we insist that default of i happen as soon as $\bar{M}_{t,t}^{(i)} < U^{(i)}$ is true. . . .

$$= \mathbb{N}\Big[\rho^{(i)} \cdot \mathbf{M} + \bar{\rho}^{(i)} M^{(i)}$$

$$\leq \Psi\left(\bar{p}_{0,t}^{(i)} - g_t^{(i)} \sqrt{t} F_t^{(i)}\left(\eta^{(i)} \cdot \mathbf{Z} + \bar{\eta}^{(i)} Z^{(i)}\right)\right)\Big], \tag{3.106}$$

where $\Psi \equiv \Phi^{-1}$ and, we recall, $\bar{\rho}^{(i)} \equiv \sqrt{1 - \rho^{(i)} \cdot \rho^{(i)}}$, $\bar{\eta}^{(i)} \equiv \sqrt{1 - \eta^{(i)} \cdot \eta^{(i)}}$.

Note that the special case $\eta \equiv 0$ (or, equivalently, $g_t \equiv 0$) corresponds to survival probabilities that are not stochastic, so that for each T,

$$\bar{M}_{t,T}^{(i)} = q_{0,T}^{(i)} = q_{t,T}^{(i)}, \quad t \leq T. \tag{3.107}$$

The model we have presented collapses in this case to the familiar static one-factor Gaussian copula model.

The reason for decomposing the uniform $U^{(i)}$ and the time-changed process X_t in terms of sums of independent terms is that by conditioning on the values of $\mathbf{M}$ and $\mathbf{Z}$, which are common to all reference names i, we can obtain the conditional joint law of default times of several reference names. Indeed, from (3.106), expanding $\Psi \equiv \Phi^{-1}$ to first order around $\bar{p}_{0,t}^{(i)}$, we end up approximating the conditional default probability for name i as

$$\mathbb{N}\left(\tau^{(i)} < t \,\big|\, \mathbf{M}, \mathbf{Z}\right)$$

$$\approx \mathbb{N}\left(\bar{\rho}^{(i)} M^{(i)} + \Psi'^{(i)} \sigma \bar{\eta}^{(i)} Z^{(i)} \leq \Psi^{(i)} - \Psi'^{(i)} \sigma \eta^{(i)} \cdot \mathbf{Z} - \rho^{(i)} \cdot \mathbf{M}\right)$$

$$= \Phi\left(\frac{\Psi^{(i)} - \Psi'^{(i)} \sigma \eta^{(i)} \cdot \mathbf{Z} - \rho^{(i)} \cdot \mathbf{M}}{[(\bar{\rho}^{(i)})^2 + \sigma^2 (\bar{\eta}^{(i)})^2 (\Psi'^{(i)})^2]^{\frac{1}{2}}}\right), \tag{3.108}$$

where $\Psi^{(i)} \equiv \Psi(\bar{p}_{0,t}^{(i)})$ indicates the inverse cumulative normal distribution function, $\Psi^{(i)\prime} \equiv \Psi'(\bar{p}_{0,t}^{(i)})$ is its derivative, easily written in terms of the Gaussian density, and where we have abbreviated $\sigma \equiv g_t F_t \sqrt{t}$.

Because, by construction, the default times $\tau^{(i)}$ are conditionally independent given $\mathbf{M}$ and $\mathbf{Z}$, the conditional loss distribution at any time t of a portfolio of names is the distribution of a sum of independent single-name loss distributions. Thus, if $L^{(i)}$ is the cumulative loss process for name i, then

$$\mathbb{N}(L_t \in (x, x + dx) | \mathbf{M}, \mathbf{Z}) = \mathbb{N}\left[\left(\sum_{i=1}^{n} L_t^{(i)}\right) \in (x, x + dx) \big| \mathbf{M}, \mathbf{Z}\right] \tag{3.109}$$

is the law of a sum of *independent* random variates. This law can be computed numerically either using Fourier inversion or, as we describe in Sect. 3.3.3 below, by recursive methods.

Having obtained the conditional loss distribution, the full unconditional loss distribution is only an integration step away, since the integral

$$\mathbb{N}(L_t \in (x, x + dx)) = \int \mathbb{N}(L_t \in (x, x + dx) \,|\, \mathbf{M} = \mathbf{m}, \mathbf{Z} = \mathbf{z}) \, \phi(d\mathbf{m}, d\mathbf{z}) \tag{3.110}$$

with respect to the gaussian densities of $\mathbf{M}$ and $\mathbf{Z}$ can be accomplished efficiently using a quadrature method.

Knowing the distribution of the portfolio loss, given in (3.110), allows us to compute prices of derivatives on the portfolio loss, in particular CDO tranches, by computing the required expectations numerically.

3.3.3 Technical Note: Recursion

The law of a sum of independent discrete random variables can be computed by a simple recursive procedure. Suppose we are given n *independent* discrete random variables $Y_1, Y_2, \ldots, Y_n$, assume that the support of Y_i is the set $\{0, 1, 2, \ldots, y_i\}$, $i = 1, 2, \ldots, n$, and let $p_i(k) = \mathbb{N}(Y_i = k)$. Consider the random variables

$$S_j = \sum_{i=1}^{j} Y_i, \quad j = 1, \ldots, n. \tag{3.111}$$

This has support in $\{0, 1, \ldots, s_j := \sum_{i=1}^{j} y_i\}$, and distribution $\{p(j, k) = \mathbb{N}(S_j = k)\}$. The probabilities $p(j, k)$ can be found by a recursive procedure, as follows:

(i) Start with $p(0, 0) = 1$, $p(0, k) = 0$, $k = 1, 2, \ldots, s_n$.
(ii) For each $j = 1, \ldots, n$, compute $p(j, k)$ for each k from 0 to s_j:

$$p(j, k) = \sum_{i=0}^{k} p(j - 1, i) p_j(k - i). \tag{3.112}$$

Note that in the case where the Y_i are Bernoulli variables with two possible values, as in the case of reference names that either default or not, all terms but two vanish in the sum (3.112).

Now, as we have seen above, conditional on the market variables $\mathbf{M}$ and $\mathbf{Z}$, the portfolio loss distribution L_t is a sum of independent distributions, so recursion can be applied once the loss distribution is suitably discretized. To do this, consider the process for the i'th reference name,

$$L_t^{(i)} := A^{-1} \bar{R}_i A_i \mathbb{1}_{\tau^{(i)} \le t}, \quad A := \sum_{i=1}^{n} A_i, \tag{3.113}$$

with $R_i = 1 - \bar{R}_i$ being the fractional recovery for i and A_i the corresponding notional at risk for name i. In terms of this the fractional portfolio loss (3.82) is

$$L_t = \sum_{i=1}^{n} L_t^{(i)}. \tag{3.114}$$

We discretize the support set of L_t, namely

$$\mathscr{A} = 0 \cup \left\{ z : z = A^{-1} \sum_{i \in \mathscr{I}} A_i \bar{R}_i \text{ for some } \mathscr{I} \subseteq \{1, 2, \ldots, n\} \right\}, \qquad (3.115)$$

by choosing a real number $\hbar$ and integers ψ_i and writing

$$A^{-1} A_i \bar{R} = \hbar \psi_i + r_i, \quad i = 1, 2, \ldots, n, \qquad (3.116)$$

with each remainder term r_i satisfying $r_i < \hbar$. In this way, the random variable L_t, with support $\mathscr{A}$, may be approximated by a discrete random variable of support

$$\{0, \hbar, 2\hbar, \ldots, K\hbar\}$$

for some sufficiently large integer K. In other words, the loss suffered by each reference name upon default, expressed as a fraction of the total portfolio notional A, is an integer multiple of a basic loss quantum, $\hbar$. In particular, the loss suffered by reference name i takes values in $\{0, \hbar \psi_i\}$ and the number of loss quanta $\hbar$ has distribution

$$p_i(k) = \begin{cases} 1 - q_{0,t}^{(i)}, & k = 0 \\ q_{0,t}^{(i)}, & k = \psi_i \\ 0, & \text{otherwise,} \end{cases} \qquad (3.117)$$

with the quantities on the right being obtainable from the observed credit spreads for reference name i.

Having discretized the support set $\mathscr{A}$, the conditional portfolio loss process

$$L_t \big| (\mathbf{M}, \mathbf{Z}) = \sum_{i=1}^{n} L_t^{(i)} \big| (\mathbf{M}, \mathbf{Z}) \qquad (3.118)$$

is a sum of discrete independent random variates, whose law can now be obtained using recursion. Clearly, the smaller the value one chooses for $\hbar$, the better will L_t be approximated by the corresponding discretized distribution, but this will necessitate a larger value of the integer K and result in longer computational time.

3.3.4 Properties of the Loss Distribution: Large Homogeneous Portfolio

The portfolio loss distribution, and the prices of CDO tranches, can be written in closed form in the special case of a one-factor Gaussian copula model, under the assumptions that the number of names in the portfolio is arbitrarily large.

To derive the closed form limiting distribution, we assume that the linear dependence parameter ρ_i in (2.72), the recovery fraction R_i, and the default probabilities

$1 - q_{0,t}^{(i)}$ are identical for all i. That is, the portfolio is homogeneous, and contains an arbitrarily large number of identical reference names.

Denoting by $\hat{L}$ the process whose value is the fraction of names that default in the portfolios, Vasicek [105] showed that the law of the time-t loss can be written as

$$\mathbb{N}\left[\hat{L}_t \leq h\right] = \Phi\left(\frac{\bar{\rho}\Phi^{-1}(h) - \Phi^{-1}(1 - q_{0,t})}{\rho}\right), \quad h \in [0, 1], \ 0 \leq \rho \leq 1, \quad (3.119)$$

where we have dropped the now-irrelevant superscripts on the survival probability $q_{0,t}^{(i)}$ and the correlation $\rho^{(i)}$.

Note that for $\rho = 0$, when defaults happen independently, the above says what we expect from the law of large numbers, namely that the proportion of losses will coincide with the probability $(= 1 - q_{0,t})$ that a single name defaults. In the limiting case $\rho \to 1$, $\hat{L}$ is a Bernoulli distribution.

More generally, for intermediate values of ρ, increasing ρ serves to skew the loss distribution to the right, with the consequence that larger mass is assigned to a larger number of defaults. Figure 3.6 shows this effect. What we infer from this is that the protection value of CDO tranches of the form $[0, k_d]$, the so called equity- or base-tranches, will decrease as ρ increases. Conversely, protecting *senior* tranches of the form $[k_a, 1]$ will cost more as ρ increases. For *mezzanine* tranches with k_a and k_d strictly different from 0 or 1, the behaviour of the protection price as a function of correlation will depend on the values of k_a, k_d and ρ.

Fig. 3.6 Inverse CDF of $\hat{L}$ for a probability of default of 5%, and for various values of correlation $\rho \in [0, 1]$

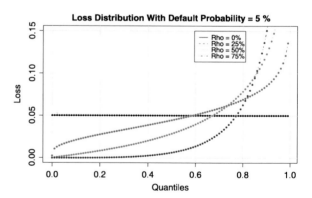

It is tedious but not to difficult to also write down the price of protection for a CDO tranche in this limiting model. Recall that a CDO tranche is the difference of two options on the portfolio loss, so we need to be able to compute expectations of the form

$$\mathbb{E}\left[\left(\hat{L}_t - \hat{K}\right)^+\right], \quad (3.120)$$

where $\hat{L}_t$, a random variable with support $[0, 1]$, is the proportion of losses suffered by the portfolio by time t, and where $\hat{K} \in [0, 1]$ is the call strike. The expectation (3.120) can be shown to equal

$$\mathbb{E}\left[\left(\hat{L}_t - \hat{K}\right)^+\right] = \Phi_2\left(-\Phi^{-1}(\hat{K}), \gamma; -\bar{\rho}\right), \tag{3.121}$$

with $\Phi_2(\cdot, \cdot; \eta)$ being the bivariate normal cumulative distribution with correlation η. Noticing that the real portfolio loss at time t will be $L_t = (1 - R)\hat{L}$, owing to the recovery fraction R, we have the price of a call option on the portfolio loss, struck at $K = (1 - R)\hat{K}$, as

$$\mathbb{E}\left[(L_t - K)^+\right] = (1 - R)\mathbb{E}\left[\left(\hat{L}_t - \hat{K}\right)^+\right], \tag{3.122}$$

whence

$$\mathbb{E}\left[(L_t - K)^+\right] = (1 - R)\Phi_2\left(-\Phi^{-1}(\hat{K}), \gamma; -\bar{\rho}\right). \tag{3.123}$$

3.3.5 Calibration of Correlation

We have seen that within the context of a Gaussian dependence model, the expression (3.110) for the portfolio loss distribution can be computed semi-analytically after conditioning on $\mathbf{M}$ and $\mathbf{Z}$.

Clearly, the loss distribution obtained will depend on the dimensionality of the factors $\mathbf{M}$ and $\mathbf{Z}$ and on the dependence parameters, $\rho^{(i)}$ and $\eta^{(i)}$ chosen for each reference entity i. This points to a way for calibrating the chosen dependence model, namely by choosing dependence parameters in such a way that model prices for chosen tranches are sufficiently close to market-observed prices. The elements of $\mathbf{M}$ and $\mathbf{Z}$ can be thought of as market factors explaining the co-dependence between reference names; for example, geographical region and credit spread level (rating).

3.4 Choice of Model

This chapter has described various models for asset and derivatives pricing, placing them in the context of the framework of Chap. 2. What we have built is a hybrid model tailored to price hybrid products whose underlying elements are the transactions in the counterparty portfolio.

The particular choice of model for each asset class is driven mainly by a balance between accuracy and simplicity, and to a certain extent by the type of products present in the portfolio. In any case, however, we need to take into account that

(i) The goal is the *pricing* and hedging of counterparty exposure, and accuracy is therefore key.
(ii) Scenario consistency, and therefore the simulation of all processes simultaneously, is essential.

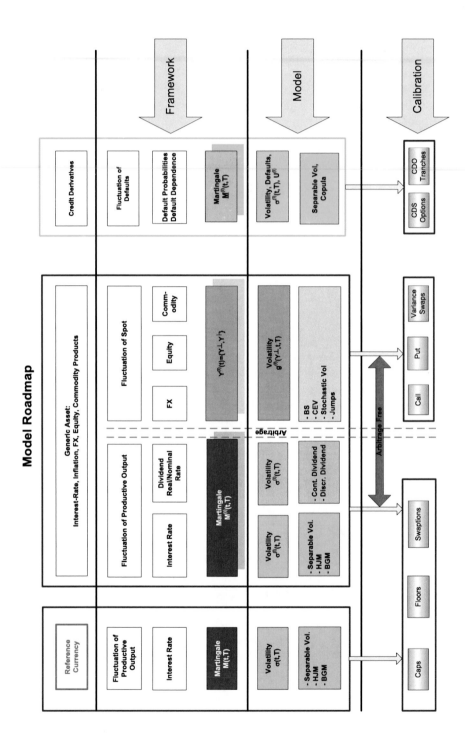

(iii) When portfolios are large, a compromise between accuracy and speed of computation needs to be found.

(iv) The interest-rate model has to be common across all asset classes.

Given the constraint of simultaneous simulation of scenarios for a large number of processes, and the valuation of portfolios of thousands of (exotic and plain vanilla) transactions, we found that one factor models with separable volatility structures (as described in Chap. 2) worked well for our purposes. In our experience, the sophistication of models used under such constraints is less important than having a setup that allows a consistent framework that can be extended in a modular way as new products require.

Chapter 4
Valuation and Sensitivities

Conceptually there are two steps in computing credit exposure: simulation followed by pricing. First, one needs to simulate scenarios from the distribution of the underlying processes that drive the price of the product concerned. Secondly, the price of this product needs to be evaluated at each time in the simulation schedule for each of the simulated scenarios.

In the previous chapters we have considered a general simulation framework and we have specified simple models used in practice. The aim of this chapter is about the second step, pricing. In simple cases pricing can be performed in closed form or semi-analytically. If a closed-form valuation, which maps scenarios to price, is not available, then the pricing step needs also to be carried out by simulation. This implies that for products with no closed-form valuation, the problem of computing price distributions entails performing simulations (for pricing) within simulations (of scenarios for the underlying processes), an approach that quickly becomes unfeasible for any reasonable simulation size.

American Monte Carlo (AMC) is a simulation technique that has been applied to the problem of *pricing* financial products with features of early exercise. As we describe later, when used in this way, the AMC method yields not just (an estimate of) the price of the product but also *the price distribution* at each point in a grid of discretized time-points. This price distribution is exactly what can be exploited for estimating the quantiles that define the level of credit exposure, so that the AMC technique can be applied to problems not just of pricing but also of credit exposure computation.

In this chapter we focus on computing exposure using AMC. We briefly describe the theoretical basis of AMC, show different algorithms used in practice, and indicate our choice of algorithm for the computation of credit exposure.

Finally, we describe different techniques to compute price sensitivities, which will be key later in Chap. 14, when discussing pricing and hedging counterparty risk.

G. Cesari et al., *Modelling, Pricing, and Hedging Counterparty Credit Exposure,*
Springer Finance, DOI 10.1007/978-3-642-04454-0_4,
© Springer-Verlag Berlin Heidelberg 2009

4.1 American Monte Carlo: Mathematical Notation and Description

When it comes to pricing financial products with very intricate payoff structures, or which depend on several underlying factors with complex dynamics, (Monte Carlo) simulation has become the tool of choice. As long as one has a means of sampling from the random distribution of the underlying drivers, there is virtually no limit to the payoff structures that can be priced. However, applying a simulation method to pricing of products with early-exercise features is not a straightforward task, the reason being that at any point in the life of such a product, the value depends on exercise decisions made at times in the future. In turn, the decision to exercise or not would depend on the perceived value of not exercising as compared to the intrinsic value upon exercise. In principle, this causes the pricing problem for an early-exercise product to mushroom into similar pricing problems on each simulated path, at each time-point considered. The number of simulations required quickly grows enough to thwart any attempt at pricing by straightforward simulation.

The technique that has now become known as *American Monte Carlo* attempts to get around this problem by performing one set of simulations and then *estimating* (rather than pricing through new simulations) at each point in time the value of not exercising (once this value is known, the task of comparing it with the intrinsic value of exercising is relatively easy). One by-product of pricing by American Monte Carlo[1] is that apart from the desired price, the method produces also samples from the price distribution at times between the pricing time and the expiry of the product. This feature makes it well-suited (with modifications) to estimating the *counterparty or market risk* posed by a particular product, such exposure being based merely on the quantiles of the product price distribution at different times.

4.1.1 Mathematical Formulation

We presented in Chap. 2 a definition of credit exposure for a generic product with early-exercise features which we have denoted by P. At the outset, the holder of P is entitled to a cashflow $X \equiv (X_t)$. We denote by T_X the maturity of X, so that $X_t = 0$ if $t > T_X$. Apart from the cashflows X, P also gives the holder the option to replace, at specific points in time, their entitlement to X with an alternative product, which we call the post-exercise portfolio, denoted by Q, and which has maturity T_Y so that Q has value zero at times after T_Y. We write

$$\mathscr{T} = \{\tau_1, \tau_2, \ldots, \tau_{n_E}\} \cup \{\infty\} \tag{4.1}$$

for the set of n_E times at which the option may be exercised to give up X in exchange for Q. If exercise happens at $\tau_E \in \mathscr{T}$, then the value provided by P until the exercise

---

[1] . . . for products both with early-exercise and without it. . . .

time is embodied in (2.3),

$$\Pi_t^{no} = N_t \mathbb{E}\left[\int_t^{\tau_E \wedge T_X} \frac{X_u}{N_u} du \,\middle|\, \mathcal{F}_t\right], \tag{4.2}$$

where the superscript on the left indicates that this is the value of the no-exercise flows X.

The optimality criterion by which the holder chooses the optimal time, τ_E^*, at which to exercise the option, will be defined shortly below. There are several possibilities for the form that the alternative holdings represented by Q may take.

(i) *Physical Settlement.* In this case, the cashflows (X_t) provided by the no-exercise portfolio Π^{no} are replaced by cashflows $(Y_t)_{0 \le t \le T_Y}$ changing also the maturity of the transaction from $T = T_X$ to $T = T_Y$. The price distribution of the trade will take values for all t in $[0, T_X \vee T_Y]$. An example of this type of product is a physically settled swaption.

(ii) *Cash Settlement* is different from physical settlement in that the net present value at time of exercise of all the flows (Y_t) is exchanged at exercise time τ_E, and the transaction then terminates. The price distribution will take values for all t in $[0, \tau_{n_E}]$. An example is a cash settled swaption.

(iii) *Intrinsic Exercise.* Here, the option holder receives the time-τ_E flow, Y_{τ_E}, *and no further cashflows*. The price distribution will take values for at most all t in $[0, \tau_{n_E}]$. An example is a Bermudan option, or a cancellable swap.

(iv) *No exercise at all.* For this case, we simply set

$$\mathcal{T} = \{\infty\} =: \mathcal{T}^{\infty}, \tag{4.3}$$

expressing the fact that exercise will never happen, and therefore that the holder of the no-exercise portfolio Π^{no} will receive flows (X_t) until expiry time T_X.

As highlighted in Chap. 2 we can then write the price distribution of product P as,

$$V_t = \begin{cases} V_t^P, & t < \tau_E^* \\ V_t^Q, & t \ge \tau_E^*. \end{cases} \tag{4.4}$$

The first element V_t^P is given by,

$$V_t^P = N_t \sup_{\tau_E \in \mathcal{T}_t} \left\{ \mathbb{E}\left[\int_t^{\tau_E \wedge T_X} \frac{X_u}{N_u} du \,\middle|\, \mathcal{F}_t\right] + \mathbb{E}\left[\frac{V_{\tau_E}^Q}{N_{\tau_E}} \,\middle|\, \mathcal{F}_t\right] \right\} \quad (t < \tau_E^*), \tag{4.5}$$

where

$$\mathcal{T}_t = \{\tau \in \mathcal{T} \mid \tau \ge t\}. \tag{4.6}$$

The second element, V_t^Q, can have different formulations depending on the type of callability. In practice, the flows X and Y ensuing from P and Q are not continuous but occur at discrete time points. For simplicity, we will nevertheless consider X

and Y to be defined for any $t \geq 0$ and set $X_t = 0$ (resp. $Y_t = 0$) if X (resp. Y) provides no cashflow at t. Note that physical and cash settlement provide the same value to the option holder. Exercising intrinsically into the cashflow (Y_t), however, provides less value, since the option holder is then not entitled to flows Y_t for $t > \tau_E$. Symbolically, if V_t^Q is the time-t value to the option holder who has exercised at $\tau_E < t$, we will have, in terms of notation introduced in Chap. 2,

$$
V_t^Q = \begin{cases} \Pi_t^{ex} = N_t \mathbb{E}[\int_t^{T_Y} \frac{Y_u}{N_u} du \mid \mathscr{F}_t], & \text{for non-intrinsic exercise} \\ \pi_t^{ex} = Y_t \mathbb{1}_{t=\tau_E}, & \text{for intrinsic exercise.} \end{cases} \tag{4.7}
$$

The value of non-intrinsic exercise is unaffected by whether settlement is physical or in cash form, save for the fact that the holder receives the flows (Y_u) in the former case, and the one-off payment Π_t^{ex} at τ_E in the latter.

At each time t, the holder of the product P attains the value V_t by choosing his exercise time $\tau_E^* \in \mathscr{T}$ so as to maximise the net present value of his cashflows.

The problem we want to consider is how to evaluate V_t. At each time $\tau_i \in \mathscr{T}$ where the option holder may potentially exercise his option, the decision whether to exercise or to continue will be based on the information observed in the economy. Formally, then, we suppose that at any time t, the information set for the model consists of a σ-algebra $\mathscr{F}_t$, part of a filtration $(\mathscr{F}_t)_{0 \leq t \leq T}$ generated by J underlying stochastic processes, say

$$
\varXi \equiv (\xi_1, \ldots, \xi_J), \tag{4.8}
$$

that drive the economy. For our model, these stochastic drivers will be, for instance, the collection of Brownian Motions, which the martingales $(M_{t,T})$ depend on. We also suppose, not unreasonably, that the holder of the option knows at t whether exercise has taken place yet, that is, $\mathscr{F}_t \supset \{\tau_E \leq t\}$. The price processes V_t^P and V_t^Q are assumed to be adapted to $(\mathscr{F}_t)$. Further, we suppose that there is a vector process

$$
\varTheta \equiv (\theta_1, \ldots, \theta_{n_{obs}}), \tag{4.9}
$$

which generates a filtration $\sigma(\varTheta)$ with $\sigma(\varTheta_t) \subseteq \mathscr{F}_t$, for each $t \in [0, T]$. Thus, for instance, θ_j could be the value process of a market instrument whose value depends on the same underlying factors, $\varXi$, in the economy as does the price process V^P. The instruments whose price processes are the θ_j are referred to as *observables*. Their importance will become clearer when we will discuss the optimal decision algorithm within the American Monte Carlo framework. The key is that at any particular time t, we will assume that the decision whether or not to exercise *will be driven by time-t values of the observables.*[2]

---

[2]Taken at face value, this assumption would seem to exclude path-dependent products; in fact it does not. It is perfectly legitimate to take as observable the price process of a path-dependent instrument, and then allow its time-t price to drive time-t exercise decisions.

Pricing of P, and estimation of the optimal exercise rule τ_E^*, is via Monte Carlo simulation; to this end we assume we have simulated realisations

$$\{\hat{\xi}_{j,k}^{(v)}\}, \quad 1 \leq j \leq J, \ 0 \leq k \leq K, \ 1 \leq v \leq n \tag{4.10}$$

of the driving factors Ξ in the economy; here, j indexes the j'th driving factor, k indexes the k'th time-point t_k in a partition, say

$$\mathscr{P} := \{0 = t_0, t_1, \ldots, t_K = T\} \tag{4.11}$$

of the time-interval $[0, T]$, and v indexes the v'th simulation out of a total of n. The $\hat{\ }$ serves to indicate a sampled value. Thus, for each t_k, we have samples of size n drawn from the distributions of each of the driving factors ξ_j. Similarly, we assume we also have simulated realisations

$$\{\hat{\theta}_{m,k}^{(v)}\}, \quad 1 \leq m \leq n_{obs}, \ 0 \leq k \leq K, \ 1 \leq v \leq n, \tag{4.12}$$

from the laws of the n_{obs} observables $\theta_1, \ldots, \theta_{n_{obs}}$.

4.1.2 Practical Examples

Let's consider some examples that illustrate how these equations should be interpreted in practice.

4.1.2.1 Non Exercisable Trades

For a transaction that does not allow early exercise, it is necessary to set,

$$\mathscr{T} \equiv \mathscr{T}^\infty = \{\infty\}. \tag{4.13}$$

In principle, we could allow $\mathscr{T}$ to be unrestricted and set $V_t^Q = -\infty$ for each $t \geq 0$.[3] Algorithmically, the first approach is neater, since then one does not need to even consider exercising at any t.

4.1.2.2 Simple Examples with Exercise

We now turn to look at how simple trades, a Bermudan put option, a cancellable swap and a European swaption, are represented in this framework.

Bermudan Put Option: Consider a 5 year contract on a stock S, which, at given dates (e.g. every year), gives the right to the holder of the option, to sell the stock at a predefined strike K.

---

[3]This is the case of long callability. In the case of short callability we write $V_t^Q = +\infty$.

The set of exercise dates $\mathcal{T}$ is defined as,

$$\mathcal{T} = \{1, 2, 3, 4, 5\} \cup \{\infty\}. \tag{4.14}$$

The no-exercise portfolio is defined as

$$X_t = 0 \quad \forall t, \tag{4.15}$$

and the exercise portfolio as,

$$Y_t = \begin{cases} K - S, & \text{if } t \in \{1, 2, \ldots, 5\} \\ 0, & \text{otherwise.} \end{cases} \tag{4.16}$$

Note that we have written $K - S$ and not $(K - S)^+$ as the optimal exercise strategy is performed by the decision algorithm.

Cancellable Swap: Consider a contract in which we start off with a 10-year swap with unit notional where we pay yearly coupons of 5% per annum and receive the 12-month libor fixed a year in advance, and in which we have the option to cancel the swap after 5 years, each year, for a fixed fee of 1%. In this case, the non-exercise payoffs (X_t) would be defined as:

$$X_t = \begin{cases} L_{t-1}[t - 1, t] - 0.05, & \text{if } t \in \{1, 2, \ldots, 10\} \\ 0, & \text{otherwise.} \end{cases} \tag{4.17}$$

The set of exercise dates $\mathcal{T}$ is defined as:

$$\mathcal{T} = \{5, 6, 7, 8, 9, 10\} \cup \{\infty\}. \tag{4.18}$$

As for the exercise portfolio, the cash flows (Y_t) are defined as

$$Y_t = \begin{cases} -0.01, & \text{if } t \in \{5, 6, \ldots, 10\} \\ 0, & \text{otherwise,} \end{cases} \tag{4.19}$$

to reflect the penalty due at time of exercise.

Given that there is a fixed one-off penalty upon exercise, the transaction has an intrinsic optionality feature, with

$$V_t^Q = \pi_t^{ex} = Y_t \mathbb{1}_{t=\tau_E}, \quad t \in \mathcal{T}. \tag{4.20}$$

Physically Settled European Swaption: Consider a contract where we have the right but not the obligation to enter in 5 years' time, into a 5-year swap of unit notional, in which we would pay yearly coupons of 5% per annum and receive the 12-month libor rate. If, in 5 years' time, we decide to exercise, then we are subject to market and counterparty risk for the remaining 5 years of existence of the swap. Otherwise, the trade terminates with no further exchange of cashflows.

In this example, the set of exercise dates is:

$$\mathcal{T} = \{5\} \cup \{\infty\}. \tag{4.21}$$

The non-exercise payoffs X are

$$X_t = 0, \quad t \geq 0. \tag{4.22}$$

For the exercise payoffs, we have

$$Y_t = \begin{cases} L_{t-1}[t-1,t] - 0.05, & \text{if } t \in \{6,7,\ldots,10\} \\ 0, & \text{otherwise.} \end{cases} \tag{4.23}$$

Choosing to exercise entails entering into the swap, so that we have non-intrinsic optionality with

$$V_t^Q = N_t \mathbb{E} \left(\sum_{t_i \geq t}^{10} \frac{Y_{t_i}}{N_{t_i}} \,\middle|\, \mathcal{F}_t \right), \quad t \geq 0. \tag{4.24}$$

In particular, we have

$$V_5^Q = A_5(S_5 - 0.05), \tag{4.25}$$

where A_t is the annuity at t of a 5-year swap with yearly coupons, and S_t is the par rate of such a swap.

Hence the optimisation program (4.5) reduces to

$$V_5 = \max\left(0, A_5(S_5 - 0.05)\right), \tag{4.26}$$

and

$$V_t = N_t \mathbb{E} \left(\frac{A_5(S_5 - 0.05)^+}{N_5} \,\middle|\, \mathcal{F}_t \right), \quad t \in [0,5]. \tag{4.27}$$

Many more types of transactions will be considered in detail at a later stage of this book.

4.1.3 Backward Induction Algorithm

There are several approaches that may be employed to compute the optimal exercise decision rule. In general a recursive procedure is used. This involves estimating at each time step t_k the expected value of *not* exercising, conditional (on not having exercised prior to time t_k and) on the time-t_k value of the observables θ_j.

The base case for the induction is the point in time where the prices of both products P and Q are trivial. This happens at time $T \equiv T_X \vee T_Y$, after which X and

Y are both identically zero by definition. At this time, we have

$$\text{(Base case)} \quad \hat{V}_T^{(\nu)} = \max\left((\hat{V}_T^P)^{(\nu)}, (V_T^Q)^{(\nu)}\right) \mathbb{1}_{T\in\mathscr{T}} + (\hat{V}_T^P)^{(\nu)}\mathbb{1}_{T\notin\mathscr{T}}$$

$$\equiv \max\left[\hat{X}_T\left(\hat{\Xi}_T^{(\nu)}\right), \hat{Y}_T\left(\hat{\Xi}_T^{(\nu)}\right)\right] \mathbb{1}_{T\in\mathscr{T}}$$

$$+ \hat{X}_T\left(\hat{\Xi}_T^{(\nu)}\right) \mathbb{1}_{T\notin\mathscr{T}}, \quad 1 \le \nu \le n, \tag{4.28}$$

where the $\hat{\ }$ indicates sampled/estimated values and where we have made explicit that the payoffs X_T and Y_T depend on the samples of the underlying driving factor $\hat{\Xi}_T$.

For valuation times $t_k < T$, we proceed inductively. Suppose that exercise has not happened prior to t_k, and write F_{t_k} for the expected value that would be gained by an agent who *does not* exercise at time t_k, but who follows the optimal strategy at times after t_k. By our assumptions, the expected value F_{t_k}, (the continuation value), is a function of the time-t_k observables Θ_{t_k}:

$$F_{t_k} \equiv F_{t_k}(\Theta_{t_k}). \tag{4.29}$$

In practice, what we have is a finite sample of size n from the distribution of the time-t_k observables Θ_{t_k}, and from this we can hope to get a sample of size n from the law of the conditional-expected non-exercise value, $F_{t_k}(\Theta_{t_k})$:

$$\frac{\hat{F}_{t_k}}{N_{t_k}} \equiv \frac{\hat{F}_{t_k}}{N_{t_k}}(\hat{\Theta}_{t_k})$$

$$:= \mathbb{E}\left[\int_{t_k}^{t_{k+1}} \frac{X_u}{N_u} du \,\Big|\, \hat{\Theta}_{t_k}\right]$$

$$+ \sup_{\tau_E \in \mathscr{T}_{t_{k+1}}} \left\{ \mathbb{E}\left(\frac{1}{N_{t_{k+1}}}\left[\int_{t_{k+1}}^{\tau_E \wedge T} \frac{N_{t_{k+1}}}{N_u} X_u du + \frac{N_{t_{k+1}}}{N_{\tau_E}} V_{\tau_E}^Q \mathbb{1}_{\tau_E \le T}\right] \Big|\, \hat{\Theta}_{t_k}\right)\right\}. \tag{4.30}$$

Again, the $\hat{\ }$ indicates sampled or estimated values.[4] The expectation in the first line in the above expression is merely the value accumulated from hesitating (at time t_k) to exercise for one more time step (until t_{k+1}). The remaining terms constitute the value to be gained from following the optimal exercise rule from time t_{k+1} onwards. To see this, note that the term in {} becomes

$$\sup_{\tau_E \in \mathscr{T}_{t_{k+1}}} \left\{ \mathbb{E}\left(\frac{1}{N_{t_{k+1}}} \mathbb{E}\left[\int_{t_{k+1}}^{\tau_E \wedge T} \frac{N_{t_{k+1}}}{N_u} X_u du + \frac{N_{t_{k+1}}}{N_{\tau_E}} V_{\tau_E}^Q \mathbb{1}_{\tau_E \le T} \,\Big|\, \hat{\Theta}_{t_{k+1}}\right] \Big|\, \hat{\Theta}_{t_k}\right)\right\}$$

---

[4]For clarity, we have suppressed the explicit ν superscript indexing simulations, but (4.30) consists of n equations. In particular, the value $\hat{F}_{t_k}$ is of course not the exact solution to the decision problem, as the supremum and expectation are obtained numerically from sampled values.

$$= \mathbb{E}\left(\frac{1}{N_{t_{k+1}}} \sup_{\tau_E \in \mathscr{T}_{t_{k+1}}} \left\{ \mathbb{E}\left[\int_{t_{k+1}}^{\tau_E \wedge T} \frac{N_{t_{k+1}}}{N_u} X_u du + \frac{N_{t_{k+1}}}{N_E} V_{\tau_E}^Q \mathbb{1}_{\tau_E \leq T} \,\Big|\, \hat{\Theta}_{t_{k+1}} \right] \right\} \,\Big|\, \hat{\Theta}_{t_k} \right).$$
$$(4.31)$$

Notice that we have used here the so-called 'tower-law' of conditional expectations. Taking the Θ_{t_k}-conditional expectation outside the sup operator is legitimate, because (and only because) Θ_{t_k} is irrelevant to the maximisation conditional on $\Theta_{t_{k+1}}$-information. Finally, invoking the definition (4.5), we translate (4.31) to

$$\mathbb{E}\left[\frac{1}{N_{t_{k+1}}} \hat{V}_{t_{k+1}}(\hat{\Theta}_{t_{k+1}}) \,\Big|\, \hat{\Theta}_{t_k} \right].$$
$$(4.32)$$

Putting everything together, we can re-write (4.30) as

$$\frac{\hat{F}_{t_k}}{N_{t_k}} \equiv \frac{\hat{F}_{t_k}}{N_{t_k}}(\hat{\Theta}_{t_k}) = \mathbb{E}\left[\int_{t_k}^{t_{k+1}} \frac{X_u}{N_u} du \,\Big|\, \hat{\Theta}_{t_k} \right] + \mathbb{E}\left[\frac{\hat{V}_{t_{k+1}}}{N_{t_{k+1}}}(\hat{\Theta}_{t_{k+1}}) \,\Big|\, \hat{\Theta}_{t_k} \right]. \quad (4.33)$$

Recall that $\hat{F}_{t_k}^{(\nu)}$ represents the ν'th estimate 'drawn' from the time-t_k value distribution of P, *conditional* on exercise not having happened prior to t_k and conditional also on it not happening at t_k.[5]

In order to obtain the time-t_k value $\hat{V}_{t_k}$ it remains to decide whether exercising at t_k results in value larger than $\hat{F}_{t_k}$, and to set

$$\hat{V}_{t_k} \equiv \hat{V}_{t_k}(\hat{\Theta}_{t_k}) = \max\{V_{t_k}^Q, \hat{F}_{t_k}\}, \quad (4.34)$$

where $V_{t_k}^Q$ (resp. $\hat{F}_{t_k}$) denotes the estimated value of exercising (resp. not exercising) at time t_k. This completes the inductive step to be made at time t_k. In the case of short optionality, where the holder of the option is the *payer* and not the receiver of the cashflows X and Y pertaining to P and Q, (4.34) becomes

$$\hat{V}_{t_k} \equiv \hat{V}_{t_k}(\hat{\Theta}_{t_k}) = \min\{V_{t_k}^Q, \hat{F}_{t_k}\}. \quad (4.35)$$

This inductive step is then repeated until all time points T_k in the partition $\mathscr{P}$ are exhausted. One fine point to mention is the following. The value process V^P is defined such that V_t^P is the value of P at time t, conditional on exercise not having happened prior to t. Now suppose that while performing the backward recursion, it is deemed optimal for the ν'th simulated path, to exercise at time t_k. It may then happen that at some later step in the recursion (and so at an *earlier* time t_j), it is also deemed optimal to exercise at t_j. By taking the exercise time τ_E to be the *earliest* time at which the recursion deems it optimal to exercise, which we assume to be the

---

[5]In (4.33) we have again suppressed the superscript indices $^{(\nu)}$ for readability.

case in the sequel, one ends up with a unique exercise strategy:

$$\hat{\tau}_E^* = \inf\left\{ t \in \mathscr{T} \mid \hat{V}_t = \sup_{\tau_E \in \mathscr{T}_t} \left[\mathbb{E}\left[\int_t^{\tau_E \wedge T_X} \frac{X_u}{N_u} du \,\middle|\, \mathscr{F}_t \right] + \mathbb{E}\left[\frac{V_{\tau_E}^Q}{N_{\tau_E}} \,\middle|\, \mathscr{F}_t \right] \right] \right\}.$$

(4.36)

In principle, computing the second conditional expectation appearing in (4.33) requires one to perform further simulations—effectively to repeat the pricing problem at each time point, and for each of the n sample values of the observables. As already mentioned, this quickly becomes an unfeasible task, and an alternative approach needs to be employed. Different approaches to Monte Carlo pricing hinge on different ways of estimating the second conditional expectation term in (4.33). There are many computational algorithms described in the literature. We move now to describing some of these approaches in details. We will then focus on our specific implementation for credit exposure computation.

4.2 AMC Estimation Algorithms

As we concluded in the previous section, it is necessary to find clever ways to estimate the conditional-expectation function

$$\Theta_{t_k} \mapsto \mathbb{E}\left[\frac{V_{t_{k+1}}}{N_{t_{k+1}}} \,\middle|\, \Theta_{t_k} \right]$$

(4.37)

appearing in (4.33), where Θ are the so called observables. This is generally done using heuristics which have shown to work well in practice. We now describe some approaches described in the literature that have been used to accomplish this, namely the Tilley [103] and the Longstaff-Schwartz [76] algorithm. The approach we employ to compute counterparty credit exposure is a modification of the regression estimation of Longstaff and Schwartz in combination with Tilley's bundling algorithm.

In all these heuristics the idea is to approximate with simple functions the continuation value and to base the decision algorithm on these approximations. This can be achieved by interpolation methods (as in the Longstaff-Schwarz algorithm) or by splitting (bundling) the domain (as in Tilley algorithm). We will see that these two approaches can be used together to obtain an efficient heuristics which compute exposure for most of the products. In the next sections we will first analyse Tilley's algorithm, as historically this was one of the first attempts to price American options.

4.2.1 Tilley's Algorithm

Tilley's algorithm was initially designed for the particular example of American options written on a single underlying stock that pays no dividends. The algorithms take into account only one observable $\Theta \equiv \theta$, and starts by first sorting the values of the observables $\{\theta^{(v)}\}$ and then classifying the samples into a chosen number of equally sized bundles. The conditional expectation (4.37) is then estimated as the sample mean for each given bundle, that is, for each sample v in some chosen bundle $\mathscr{B}$, Tilley sets

$$\frac{\hat{F}_{t_k}^{(v)}}{N_{t_k}^{(v)}} = \frac{1}{B} \sum_{v \in \mathscr{B}} \frac{\hat{V}_{t_{k+1}}^{(v)}}{N_{t_{k+1}}^{(v)}}, \tag{4.38}$$

to be the estimated value of not exercising at time t_k, with B being the number of elements in the bundle $\mathscr{B}$. Thus, Tilley's algorithm uses the information from θ to partition the estimation set into different bundles. As we will see in the next section, an alternative approach is to regress the continuation value against the observables. Thus, Tilley's method is akin to fitting a piecewise-linear function to a non-linear data set. We will see that the Longstaff-Schwartz approach fits a non-linear function to the entire data set. Notice also that there is more than one way of choosing the first path v^* for which exercise is optimal; in his original paper, Tilley proposes one particular rule to choose a unique v^*.

4.2.2 Longstaff-Schwartz Regression

Longstaff and Schwartz [76] put forward an algorithm that models the continuation value $\hat{F}_{t_k}^{(v)}$ at each time t_k, as a regression on the time-t_k value Θ_{t_k} of the chosen observable of the discounted values computed at time t_{k+1}. The idea is that, at time t_k, the sample conditional expected value of not exercising (that is the continuation value $\hat{F}_{t_k}(\hat{\Theta}_{t_k})$ in (4.31)) can be expressed as a linear combination of *basis functions* of the time-t_k observables. For this reason this algorithm is also called *regression* algorithm. These basis functions are generally polynomials. Thus,

$$F_{t_k}(\Theta_{t_k}) := \sum_j a_j L_j(\Theta_{t_k}). \tag{4.39}$$

Here it is supposed that the conditional expectation function is in a space that is spanned by the basis functions $\{L_j\}$, $j = 1, 2, \ldots$. Notice that in general, Θ is an n_{obs}-vector so that each L_j maps $\mathbb{R}^{n_{obs}}$ to $\mathbb{R}$. The fit coefficients a_j are estimated through regression, as described below.

For the Longstaff-Schwartz algorithm, the choice that needs to be made is what number of basis functions, μ say, one should use. Then, at the k'th time step, the regression coefficients (a_j) are estimated by regressing the n sample discounted values

$$\frac{N_{t_k}^{(v)}}{N_{t_{k+1}}^{(v)}} \hat{V}_{t_{k+1}}^{(v)}, \quad v = 1, 2, \ldots, n, \tag{4.40}$$

on the first μ basis functions evaluated on the sampled observables,

$$L_1(\hat{\Theta}_{t_k}^{(v)}), \ldots, L_\mu(\hat{\Theta}_{t_k}^{(v)}), \quad v = 1, 2, \ldots, n. \tag{4.41}$$

The regression carried out by the authors used only those sample values v which are in the money. In particular, for the American put example considered by Longstaff and Schwartz, the time-t_k regression used only those sample values $S_{t_k}^{(v)} < K$ of the stock price for which exercise would yield non-zero payoff.

4.2.3 Biases of Estimates

The Tilley estimation algorithm just described yields, of course, only an approximate solution to problem of pricing a product P with Bermudan exercise[6] allowed at any of the time points $\tau_k \in \mathscr{T}$. The end result obtained is influenced by both upward and downward biases that arise from the necessity to use a finite number of simulations (see also Hyer [65]).

(i) *Granularity bias* is a bias that arises because bundles are 'too big'. The cause of this bias is that the conditional-expected value of not exercising is identical for *all* paths in the same bundle. Because the same non-exercise value is used for each path in a given bundle, the estimated exercise rule for paths in that bundle will be sub-optimal, causing a downward bias in estimated price. Granularity bias would be eliminated if each path were itself a bundle.

(ii) *Small-sample bias* is that arising because bundles are 'too small'. The fact that each bundle contains only a small finite number of paths causes the algorithm to work out a sub-optimal exercise rule, again causing a downward bias in price.

(iii) *Look-back bias* is that arising because the same set of N paths is used in estimating the optimal decision rule as is used to compute the value yielded by that rule. Tilley showed by example that this type of bias results in an upward bias in the estimated price; he did this by comparing the value estimated by his algorithm to the value one obtains when using the *same* set of paths, *but* employing the exact known optimal exercise strategy.

The presence of an upward bias may at first seem contradictory, as by definition no estimated exercise strategy can dominate the optimal one. Consider, however, a situation in which each bundle consists of a single path, and pick some path (equivalently, bundle) η. Then, the backward induction algorithm would compute

$$\hat{V}_T^{(\eta)} = \hat{Y}_T^{(\eta)}$$

---

[6]If the product P allows American exercise, the need to discretize forces us to model a corresponding product with Bermudan exercise features; we will not discuss inaccuracies arising due to this. What we refer to in this paragraph, rather, are the discrepancies between the true and computed values for the problem of pricing a product P with *Bermudan* exercise features.

$$\hat{F}_{t_k}^{(\eta)} = \frac{N_{t_k}}{N_{t_{k+1}}} \mathbb{E}[\hat{V}_{t_{k+1}}^{(\eta)} | \Theta_{t_k}], \quad k = K, K-1, \ldots, 0$$

$$\hat{V}_{t_k}^{(\eta)} = \max\{V_{t_k}^Q, \hat{F}_{t_k}^{(\eta)}\}. \tag{4.42}$$

Thus, the exercise rule for the path η would depend on the simulated values of that path alone. The time-zero price would then be the expected value over all paths of the terminal payoff on each path, $\hat{V}_T^{(\eta)}$, discounted back and replaced by the exercise value whenever the latter is larger.

Mathematically, the look-back bias arises because of the convexity of the max operator and Jensen's inequality—the price is estimated as the expectation of maxima of per-path future values, rather than (correctly) as the maximum of expected values over all possible choices of exercise strategy.

4.2.4 An AMC Algorithm to Compute Credit Exposure

The drawback of Tilley's approach is in its use of a single observable to carry out bundling. In practice, transactions will depend on several underlying variables, each of which should be considered in partitioning the sampled paths. On the other hand, the Longstaff and Schwartz method computes a regression to estimate the continuation value $\hat{F}_{t_k}$, but does not employ bundling. A possible extension is a combination of modified algorithms based on these two basic ideas (see also Hyer [66]). Furthermore, motivated by Sect. 4.2.3, we introduce a bias correction device to improve the quality of estimates $\hat{F}_{t_k}$.

4.2.4.1 Recursive Bundling

Tilley defines bundles by classifying the number of simulation paths according to the level of a single observable. We generalise this notion by bundling recursively on all the n_{obs} observables in the process Θ. To see how this can be accomplished, consider a particular time point t_k. Choosing integers

$$m_1, m_2, \ldots, m_{n_{obs}}, \tag{4.43}$$

we start by classifying paths into m_1 bundles based on the level of the observable θ_{1,t_k}. Each of these is then subdivided into m_2 bundles using the level of the observable θ_{2,t_k}. The procedure is then repeated for all observables, resulting in a total of $m_1 m_2 \ldots m_{n_{obs}}$ bundles.

The reason for performing bundling is to classify the simulation paths into subsets such that for any two paths in a particular bundle, all observables have similar values. For observables with continuous distributions, this is accomplished well enough by the above recipe, which allocates paths equally across bundles. However, for a discontinuous observable such as, for instance, that corresponding to the default indicator of a reference credit, equal allocation can result in two paths with the same values for an observable being in different bundles. To prevent this, we perform an additional clustering check, and shift paths from one bundle to another, if by doing so the distance between the particular path and its closest neighbour in the bundle is reduced.

4.2.4.2 Regression

We adapt the Longstaff Schwartz method by performing regression on each bundle. Thus, for a typical bundle, $\mathscr{B}$, say, we find parameters a and b that fit the model

$$\hat{V}_{t_k} = \mathbb{E}\left(\frac{N_{t_k}}{N_{t_{k+1}}}\,\hat{V}_{t_{k+1}}\,\bigg|\,\mathscr{F}_{t_k}\right) = \sum_{j=0}^{J} a_j \theta_{1,t_k}^j + \sum_{l=0}^{L} b_l \theta_{l+1,t_k} + \varepsilon_{t_k}, \qquad (4.44)$$

where $\varepsilon_{t_k} \sim N(0, \sigma_{t_k}^2)$, and ε_{t_j} is independent of any ε_{t_k} for $j \neq k$. What we are doing here, then, is to estimate the conditional expectation as a polynomial of order J in the first observable and a linear function in the remaining $L+1$ observables.

Note that cashflows paid at time t_k are not included in the regression step, as these are known quantities.

4.2.4.3 Bias Correction

We pointed out, in Sect. 4.2.3, that several biases arise in the Tilley estimation algorithm. Fries [46] describes how the look-back bias (which he refers to as foresight bias) can be removed analytically.

Consider our regression model (4.44). For each t_k, this models the conditional expectation function

$$\Theta_{t_k} \mapsto \mathbb{E}\left[\frac{V_{t_{k+1}}}{N_{t_{k+1}}}\,\bigg|\,\Theta_{t_k}\right] \qquad (4.45)$$

as

$$\Theta_{t_k} \mapsto f(\Theta_{t_k}). \qquad (4.46)$$

Suppose that $\varepsilon \sim N(0, \sigma^2)$ is the Monte Carlo error in the estimator f, so that

$$f(\Theta_{t_k}) = \mathbb{E}\left[\frac{V_{t_{k+1}}}{N_{t_{k+1}}}\,\bigg|\,\Theta_{t_k}\right] + \varepsilon. \qquad (4.47)$$

If K is the intrinsic value of exercising at time t_k, then look-back bias arises from Jensen's inequality, because the value

$$\mathbb{E}[\max(K, f(\Theta_{t_k}))] \qquad (4.48)$$

of the exercise strategy computed by the algorithm differs from the theoretical value

$$\max\left(K, \mathbb{E}\left[\frac{V_{t_{k+1}}}{N_{t_{k+1}}}\,\middle|\,\Theta_{t_k}\right]\right). \tag{4.49}$$

Analytical removal of the bias is possible once we note that for any real a, b, and for $\varepsilon \sim N(0, \sigma^2)$, we have

$$\mathbb{E}\big[\max(a, b + \varepsilon)\big] = \sigma\phi(\eta) + \eta\sigma\,\Phi(\eta) + a, \tag{4.50}$$

with $\eta \equiv (b - a)/\sigma$.

Similarly, for short optionality, the bias correction can be removed using:

$$\mathbb{E}\big[\min(a, b + \varepsilon)\big] = -\mathbb{E}\big[\max(-a, -b - \varepsilon)\big]. \tag{4.51}$$

4.3 Post-Processing of the Price Distribution

The inductive procedure outlined in Sect. 4.1.3 provides, at each valuation / decision time t_k, an (estimate) of whether exercise at t_k is optimal *if it has not happened prior to t_k*. In order to obtain the correct estimate for the value of the problem, then, one needs to locate the *earliest* time τ_E^* at which exercise has been estimated to be optimal to the alternative of continuing. Symbolically, for each path v, we set

$$\tau_E^{(v)*} = \inf\left\{t_k \in \mathscr{T} \,\middle|\, \hat{V}_{t_k}^{(v)} = \max\left(\hat{F}_{t_k}, V_{t_k}^Q\right) = V_{t_k}^Q\right\}. \tag{4.52}$$

Following this, we then have, for each $t_k \geq \tau_E^*$,

$$\hat{V}_{t_k} = \begin{cases} V_{t_k}^Q \mathbb{1}_{t_k=\tau_E^*}, & \text{for non-intrinsic cash settlement} \\ V_{t_k}^Q, & \text{for non-intrinsic physical settlement} \\ \pi_{t_k} = Y_{t_k}\mathbb{1}_{t_k=\tau_E^*}, & \text{for intrinsic exercise,} \end{cases} \tag{4.53}$$

where we have, for ease of notation, suppressed the superscript $^{(v)}$ indexing paths.

4.4 Practical Examples Revisited

We revisit here our two illustrative examples from Sect. 4.1.2 in order to show how our algorithm would work in a concrete setting.

Cancellable Swap: In this example, we considered a cancellable swap where the holder has the option, at each $\tau_k \in \mathscr{T}$, to exit the transaction for a fixed penalty. Clearly, then, at each τ_k, it would be rational to exercise the option if the estimated value of continuing to receive the swap payments is less than the reward $Y_{\tau_k} = -0.01$ suffered from cancellation.

The observables Θ, on which the estimation of the conditional expectation is based, should be chosen while keeping in mind the underlying variable in the trade. In this case, quantities such as the Libor rate or the fair swap rate are valid observables. Using these, the backward induction would, at each time τ_k estimate the value of the cancellable swap assuming exercise has not previously taken place, and would decide to cancel the swap if its value is lower than the exercise penalty.

Physically Settled European Swaption: Recall that for the swaption example, the only allowable exercise time was $t_k = 5$. If we assume that the value of the swap annuity, A_t, and the fair swap rate, S_t, are known at each t, then the decision needs to be made only at $t_k = 5$, whether entering into the swap (with value $A_5(S_5 - 0.05)$), has value that exceeds the zero-value of the strategy of allowing the swaption to expire unexercised.

At times t prior to $t_k = 5$, the swaption has continuation value $\hat{F}_t$ which will be estimated using regression and the chosen observables. Again in this case, since the trade depends on the fair value swap rate S, it makes sense to choose this and / or related quantities as the observables Θ in the estimation.

4.5 Computing Price Sensitivities

No valuation framework would be complete without the capability of computing the sensitivities of a trade to its underlying risk drivers. The usual definition of a price sensitivity is the partial derivative of the price with respect to a given risk driver (keeping all remaining risk drivers constant).

Let $(\xi^{(i)})$ denote the stochastic processes driving the pricing of a product and for which we want to evaluate the sensitivities. Examples of risk drivers are the stock price, the FX rate, the swap rate, the zero rate, or the volatility level. If $V_0 = V(\Xi_0) = V(\xi^{(i)}(0), \ldots, \xi^{(n)}(0))$ is the time-zero price of the transaction as an explicit function of the underlying risk drivers, then it is usual to define *delta* of the trade with respect to $\xi^{(i)}$ as

$$\Delta^{(i)} := \left. \frac{\partial V}{\partial \xi^{(i)}} \right|_{\Xi = \Xi_0}, \tag{4.54}$$

the partial derivative evaluated at the time-zero value of the risk drivers.

Similarly, the *gamma* of the trade with respect to $\xi^{(i)}$ and the *cross-gamma* of the trade with respect to $\xi^{(i)}$ and $\xi^{(j)}$ are

$$\Gamma^{(i)} := \left. \frac{\partial^2 V}{\partial (\xi^{(i)})^2} \right|_{\Xi = \Xi_0} \tag{4.55}$$

and

$$\Gamma^{(i,j)} := \left. \frac{\partial^2 V}{\partial \xi^{(j)} \partial \xi^{(j)}} \right|_{\Xi = \Xi_0}. \tag{4.56}$$

4.5.1 The Classical Approach

In the banking industry, the standard way of computing price sensitivities is via a finite-difference approximation,[7] that is, modifying the set of market data to produce a small change in the value of the risk driver of interest and then re-valuing the trade.[8]

There are two obvious problems which can arise from this methodology.

(i) *Computational Speed.* Computing finite differences requires a full revaluation of the price distribution for each change in the market data. Thus, the first and second derivative of a product with respect to only one risk driver require already tripling the computation effort.

(ii) *Flexibility.* It may not be possible to perturb a chosen risk driver while keeping all other drivers constant. An example is the swap rate which is the combination of several stochastic quantities. This constrains the set of drivers for which it is possible to specify sensitivities.

4.5.2 Price Sensitivities through Regression

Within our framework, we can estimate price sensitivities at almost no extra computational cost. The idea is to regress incremental change in values of the price distribution against corresponding incremental values of the risk driver of interest.

In detail, choose $\varepsilon > 0$ to be small and write

$$\Delta V = V_\varepsilon - V_0 \equiv V(\Xi(\varepsilon)) - V(\Xi(0)). \tag{4.57}$$

Assuming that a sample of the price distribution at time ε is available, (which we can ensure at valuation stage), and choosing some $\xi^{(i)}$ to be the risk-driver of interest, we write ΔV as a polynomial in $\Delta \xi^{(i)} = \xi^{(i)}(\varepsilon) - \xi^{(i)}(0)$,

$$\Delta V = \sum_{k=0}^{m} a_k \left(\Delta \xi^{(i)} \right)^k. \tag{4.58}$$

Regression gives us estimates of the weights a_k, which, in turn, enable us to compute the k partial derivatives

$$\frac{d^k V}{d(\xi^{(i)})^k} = k! a_k. \tag{4.59}$$

It is important to note that this is the total derivative of the transaction value with respect to the chosen $\xi^{(i)}$, and that $\xi^{(i)}$ may itself be correlated to other risk factors

---

[7] There are several ways to compute sensitivities. For a survey see for example the book by Glassermann [50].

[8] Practitioners often refer to this technique as *bumping*.

that affect V. Obtaining the true partial derivative of V with respect to a chosen factor, eliminating the effect of correlation between the risk factors themselves, is the subject of the next section.

By way of illustration, Table 4.1 compares the prices of European options and their sensitivities—obtained through AMC with 10,000 paths—to their analytical Black-Scholes values. The errors are typically less than the difference in price resulting from a change of one volatility point.

Table 4.1 Performance of AMC for price and Greeks on one-year European options. $S = 100$, $r = 2.95\%$, $\sigma = 20\%$

Type	BSPrice	AMCPrice	BSDelta	AMCDelta	BSGamma	AMCGamma
Call @ 105	7.106	7.100	0.501	0.494	0.0199	0.0198
Call @ 100	9.388	9.386	0.597	0.588	0.0193	0.0193
Call @ 95	12.151	12.153	0.693	0.682	0.0176	0.0176
Put @ 95	4.389	4.366	−0.307	−0.301	0.0176	0.0167
Put @ 100	6.481	6.454	−0.402	−0.394	0.0193	0.0184
Put @ 105	9.054	9.024	−0.499	−0.489	0.0199	0.0189

4.5.3 Removing Correlation

As we saw above, sensitivities to a risk driver $\xi^{(i)}$ computed through regression implicitly contain sensitivity also to other risk drivers that are correlated to $\xi^{(i)}$.

While this is useful in the case where we compute only one sensitivity, since it can give more information on the risk of the trade (and therefore better hedges), it is an undesirable effect if we want to use the sensitivities as a tool to explain daily changes of profit and loss (P&L) of a business. It also becomes undesirable as soon as we are interested in sensitivities to more than one risk driver.

With some assumptions it is possible, however, to remove this correlation effect and to produce *decorrelated* sensitivities, that is, sensitivities with respect to a chosen risk driver while keeping all remaining risk drivers constant. To this end, suppose we have m simulated values of each of the n risk drivers $\xi^{(i)}$, $i = 1, \ldots, n$, and let $\hat{X}$ be the $m \times n$ matrix defined by[9]

$$\hat{X}_{j,i} = \widehat{\Delta\xi^{(i)}_j}, \quad i = 1, \ldots, n, \ j = 1, \ldots, m, \tag{4.60}$$

where the $\hat{\ }$ indicates a sampled (simulated) value and the superscript indicates the sample index. By this, the i'th column of $\hat{X}$ is a sample of size m from the distribution of $\Delta\xi^{(i)}$, the incremental change in the i'th risk driver. In what follows below,

---

[9]See Sect. 2.3.1 to clarify notation.

we make the reasonably accurate assumption that all incremental moves X in the risk drivers (which happen over a very short time period ε) are normally distributed with mean zero.[10]

Now consider the estimation of a linear model for $Y \equiv \Delta V$ in terms of the risk drivers $\boldsymbol{\Xi}$,

$$\hat{\mathbf{Y}} = \hat{\mathbf{X}}\boldsymbol{\alpha} + \hat{\boldsymbol{\varepsilon}}, \quad (\hat{\boldsymbol{\varepsilon}} \sim N(\mathbf{0}, \boldsymbol{\Sigma})), \tag{4.61}$$

where $\hat{\mathbf{Y}}$ is a vector containing a sample of size m from the distribution of ΔV. The least-squares estimate of $\boldsymbol{\alpha}$ in this model is given by

$$\boldsymbol{\alpha} = (\hat{\mathbf{X}}^{\mathsf{T}}\hat{\mathbf{X}})^{-1}\hat{\mathbf{X}}^{\mathsf{T}}\hat{\mathbf{Y}}. \tag{4.62}$$

Compare this to the corresponding estimate we would get for $\boldsymbol{\alpha}$ if a univariate regression were to be performed on just one of the risk drivers $\xi^{(i)}$, namely

$$\tilde{\alpha}_j = (\hat{\mathbf{X}}_j^{\mathsf{T}}\hat{\mathbf{X}}_j)^{-1}\hat{\mathbf{X}}_j^{\mathsf{T}}\hat{\mathbf{Y}}, \quad j = 1, \ldots, n, \tag{4.63}$$

where $\hat{\mathbf{X}}_j$ is the j'th column of $\hat{\mathbf{X}}$. Putting together the estimates for $\tilde{\boldsymbol{\alpha}} = (\tilde{\alpha}_1, \ldots, \tilde{\alpha}_n)$, we have

$$\tilde{\boldsymbol{\alpha}} = \mathbf{D}^{-1}\hat{\mathbf{X}}^{\mathsf{T}}\hat{\mathbf{Y}}, \tag{4.64}$$

where $\mathbf{D}$ is a diagonal matrix whose (i, i)'th entry is a multiple of an unbiased estimator of the variance of $\xi^{(i)}$, namely

$$D_{i,i} = \hat{\mathbf{X}}_i \cdot \hat{\mathbf{X}}_i =: (m - 1)\widehat{\mathrm{Var}}(\xi^{(i)}). \tag{4.65}$$

All this means that $\tilde{\boldsymbol{\alpha}}$ and $\boldsymbol{\alpha}$ are related by

$$\tilde{\boldsymbol{\alpha}} = \mathbf{D}^{-1}(\hat{\mathbf{X}}^{\mathsf{T}}\hat{\mathbf{X}})\boldsymbol{\alpha}. \tag{4.66}$$

In the above note that $\hat{\mathbf{X}}^{\mathsf{T}}\hat{\mathbf{X}}$ has entries

$$[\hat{\mathbf{X}}^{\mathsf{T}}\hat{\mathbf{X}}]_{i,j} = \hat{\mathbf{X}}_i \cdot \hat{\mathbf{X}}_j =: (m - 1)\widehat{\mathrm{Cov}}(\xi^{(i)}, \xi^{(j)}), \tag{4.67}$$

and is therefore simply a multiple of an unbiased estimate of the covariance matrix of $\boldsymbol{\Xi}$. In particular, the diagonal elements of $\mathbf{D}$ and $\hat{\mathbf{X}}^{\mathsf{T}}\hat{\mathbf{X}}$ are equal.

The significance of $\boldsymbol{\alpha}$ and $\tilde{\boldsymbol{\alpha}}$, as estimates derived from two different linear models for Y, is that the components of $\boldsymbol{\alpha}$ represent partial derivatives to the $\xi^{(i)}$ while those of $\tilde{\boldsymbol{\alpha}}$, obtained by regression on $\xi^{(i)}$ alone, represent full derivatives with respect to $\xi^{(i)}$. The expression (4.66) therefore allows us to obtain the true partial derivatives in terms of the correlated sensitivities computed from regression.

To evaluate the accuracy of this methodology and the effect of correlation, consider the following simple example: suppose that we enter into a trade which pays

---

[10]Recall that we have defined our framework so that all risk drivers are derived from simulated Brownian Motions.

us 100 mEUR on October 20, 2015 and in which we pay 100 mGBP on October 20, 2010. Table 4.2 summarises the deltas[11] (in terms of percentage point moves) obtained by bumping, regression and decorrelation, expressed in USD, along with the difference in computational time required for the three methods.

We can see that correlated deltas are significantly different from the de-correlated ones. This example also suggests that the accuracy of the decorrelated deltas compared with the deltas obtained via numerical differentiation is acceptable, given the benefit of the higher computational speed achieved.

In practice, before using this methodology on a large portfolio, it is necessary to carefully assess its accuracy for different types of products and over different time horizons. One possibility is to predict portfolio movements using sensitivities, and compare the results either with historical valuations, or with full revaluation of the portfolio.[12]

Table 4.2 Comparison between deltas computed by finite difference and by regression using correlated and decorrelated method. Simulation has been experimented on a desktop Intel Core 2.13 GHz machine

Risk Driver	Delta		
	Finite Difference	Correlated	Decorrelated
EUR Rates	−6,933,027	−1,989,992	−7,177,312
EURUSD FX Rate	824,200	278,850	892,464
GBP Rates	2,128,061	−1,881,813	2,355,130
GBPUSD FX Rate	−973,972	−620,948	−1,046,550
Computation Time	4.27 s	1.06 s	1.09 s

4.6 Extensions

American Monte Carlo valuation techniques have been analysed in various papers. We have already mentioned the Longstaff-Schwartz [76] and the Tilley [103] algorithm. Haugh & Kogan [59] and Rogers [91] introduced a dual method for pricing American options, providing an upper bound for the price of the option. Andersen & Broadie [2], Broadie & Glasserman [20], and Broadie & Cao [19] further develop this methodology obtaining both an upper and lower bound for the Bermudan option price.

---

[11] For interest rate deltas, we have considered a parallel-shift type of delta.

[12] In the financial industry this procedure is often called P&L explain.

Part II
Architecture and Implementation

Chapter 5
Computational Framework

In Part I we described a general framework that allows the specification of models for different asset classes, and we showed how the AMC valuation technique gives the possibility of computing price distributions, hence estimating counterparty exposure.

Our goal is now to show how this *mathematical* framework can be naturally translated into a *computational* framework that will enable the computation of exposure in a systematic way for all types of products across the asset classes we provided models for. The basic ideas we highlight in this chapter will lead to the description of a basic software architecture, which can be used to address typical integration problems that large financial institutions face. The motivation for many of the challenges we consider in this and the following chapters, as well as many of the choices we take, will become clearer in Part IV, where the computation, controlling, and hedging of exposure, will be done at counterparty and not just at trade level.

5.1 AMC Implementation and Trade Representation

Recall the basic principles of the AMC algorithm. Products are described by defining the cashflows X_t the holder a product P is entitled to, the cashflows Y_t of an alternative product Q, which could replace P on a predefined set of exercise dates $\mathscr{T}$ (which could include $\mathscr{T}^{\infty}$), and the exercise strategy τ_E. In practice the cashflows X and Y will be defined on a set of dates $\mathscr{S} = \{T_1, T_2, \ldots, T_n\}$, called *event dates*, which in general contain $\mathscr{T}$. With this information we can fully describe the product and then, using the AMC backward induction process, estimate prices along different scenario paths.

To proceed a step further, at this point one remark is key. When we use the AMC algorithm we do not need to explicitly define products. A product is implicitly described via

  (i) the cashflows X, paid before exercise, and the cashflows Y of the exercise portfolio, paid on the event dates $\mathscr{S}$,

G. Cesari et al., *Modelling, Pricing, and Hedging Counterparty Credit Exposure,*
Springer Finance, DOI 10.1007/978-3-642-04454-0_5,
© Springer-Verlag Berlin Heidelberg 2009

(ii) the set of exercise dates $\mathcal{T}$, and
(iii) the type of exercise.

If we find a way to describe and process this information systematically, we have provided a computational framework that gives the capability of computing exposure for all types of products without knowing a priori the product type.[1]

Consider the cashflows X and Y defined on $\mathcal{S}$. In general they are expressed as functions of financial quantities, such as Libor rates, FX values, or stock prices. These functions can be relatively complex, depending on the nature of the product. To be able to describe them we need to have basic building blocks, which we can combine to obtain the desired results. These building blocks, which we call *statistics*, are functions of the simulated paths and can be combined together using mathematical functions. We could have for example a statistic that extracts the Libor rate observed at a certain date t over a time interval (T_1, T_2), or a statistic that computes the swap rate, or the average stock price over a period of time, and functions that e.g. compare values, provide their max or min, and multiply or divide them.

With the description of cashflows X and Y, and with a mechanism to choose at predefined dates $\mathcal{T}$ if we want to exercise or continue, we can now compute counterparty exposure for generic products. From a computational point of view we simply need to compare values computed via the backward induction steps with predefined cashflows.

We have turned the problem from defining individual products, to describing *features* of products.

5.1.1 Examples

To clarify these points consider again the examples we have shown in the previous chapter.

Non-exercisable products: To define products that cannot be exercised we need to describe only X. As a concrete example consider an up-and-out option on a stock struck at K and knocking out at H. This product tracks a stock price and its running maximum. If its value does not exceed the barrier H it will pay at maturity the difference between the stock price and the strike K, provided that its value is positive. Mathematically we can describe its payoff as

$$X_T = (S_T - K)^+ \mathbb{1}_{\max_{u \in [0,T]} S_u \leq H}. \tag{5.1}$$

To describe $(S - K)^+$ we need a statistic that, given a Brownian path of the stock, extracts the value of the stock price S at maturity T, and two mathematical functions, one performing differences and one finding the max between values. To describe the second part of the payoff (related to the barrier), we need a statistic that computes

---

[1] ...provided that the underlying model is appropriate to describe the features of that class of products.

the extremum value of the stock price between 0 and t, and an indicator function that returns zero or one depending on the extremum and the barrier value.

If we implement these statistics we can compute the payoff of this product at maturity for each underlying path of the stock, and then rely on AMC to estimate intermediate prices from maturity to trade inception, and thus to compute counterparty credit exposure.

Other common examples of non-exercisable products are vanilla swaps. In this case the value of the cashflows X defined at coupon dates T_i are $\alpha(L_{t,T_{i-1},T_i} - c)$, with α being the day count fraction, c the fixed rate, and L the statistic that provides the Libor rate paid over a time interval. The computational mechanism is the same, using a statistic that provides the Libor rate, describes the payoff at each payment date and then uses AMC to evaluate intermediate prices.

Cancellable Swaps: As we have seen, cancellable swaps are products with *intrinsic* optionality. We need to describe at each t in $\mathcal{T}$ not only X, but also Y. When we exercise, however, we only need to replace the value of X with Y.

Physically Settled European Swaptions: In a physically-settled European swaption, the event dates T_i are the union of the swaption exercise date and swap coupon dates. X is equal to zero and Y is equal to the swap cashflows. The optionality is with physical settlement, meaning that upon exercise we replace the cashflows X with Y, entering into a swap.

5.1.2 Expression Trees

We have seen that products can be described via their payoff cashflows X and Y and that their description can be done systematically using statistics as building blocks. We can now consider how to have a mechanism to produce computing code that generates the expressions needed to describe payoffs.

The standard way to build and evaluate expressions is to use trees that are generated according to predefined rules, i.e. a predefined grammar. Consider as an example the expression tree of the up-and-out option as described in (5.1). At maturity, i.e. at the event date, we will have to evaluate an expression tree as shown in Fig. 5.1.

5.2 A Portfolio Aggregation Language

We can define products via their expression tree and then, using this information, drive the AMC algorithm to compute price distributions. In Chap. 6 we describe some basic implementation principles. Our goal now is to add an abstraction layer in order to

(i) easily compute exposure of trades that usually are described via termsheets,
(ii) de-couple trade description from implementation of the analytics, and

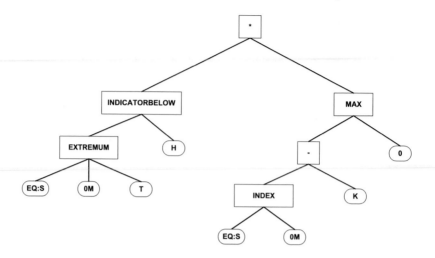

Fig. 5.1 An expression tree for an up-and-out option. We have used the following statistics (see Chap. 6 for more details): INDEX to extract stock information, EXTREMUM to obtain the extremum value of a path, INDICATORBELOW to compute the indicator function. In addition we have used mathematical functions to perform differences and multiplication between values and to compute the max between values. The hierarchy of operators is given by the grammar that generates the tree

(iii) bring trades from existing booking systems into a single unified booking representation.

The technical solution for these requirements is the definition of a programming language to describe products, which acts as an interface between statistics and analytics. As the main goal is to allow a portfolio view, we have called this language Portfolio Aggregation Language (PAL).

By defining an appropriate syntax and grammar, and by using a lexer and parser, we can then generate expression trees, which in turn will call the statistics defined in the analytics. There are many tools available in the market to automatically generate parsers from given grammars. Classical examples are Lex and Yacc [74], Flex and Bison [73], or ANTLR [84].

PAL is designed with two competing goals in mind. It has to be

(i) Simple enough to describe different types of trades in a clear and concise way. In other words, the syntax has to allow trade description in a way that is close to business language.
(ii) Flexible enough to accommodate various levels of trade complexity and allow translation from other different booking systems across the firm into one single language.

To respond to these requirements we have designed PAL with the following main technical features.

(i) It has the typical declarative statement of a procedural language. For example it is possible to define numerical or logical (boolean) variables, arithmetic operations between them (with the usual precedence rules), and loops.

(ii) There are some predefined types such as vectors or date schedules.

(iii) There are custom and built-in functions: examples are Exp, Max, Log.

(iv) It has some object-oriented capabilities to define, for example objects of type 'Instrument' (this allows the possibility of having analytical pricing for selected products).

(v) It has some typical financial construct. For example you can say that you will Pay or Receive a certain amount at certain dates.

(vi) It defines a context that specifies how computation is performed. For example it is possible to define the quoting currency or the fact that a counterparty is collateralised.

The following examples clarify these concepts.

5.2.1 PAL Examples

The code snippet below shows how a simple interest-rate swap can be described in a succinct way.

Table 5.1 Vanilla swap

```
Schedule = From 2009/09/30 to 2019/06/30 every 3 months;
Notional = 100 mm EUR;
DcFr = DCF(now-3m, now, "ACT/ACT"); // Day Count Fraction
Receive Notional * (ir:eur3m on Now - 3m)* DcFr on Schedule;
Pay Notional * 3.5%* DcFr on Schedule;
```

Even if there are some elements typical of programming languages, the syntax is close enough to a typical termsheet description: we can define a schedule using dates, we can specify what parties A and B respectively pay and receive, and we can use typical business terminology. The same swap in arrears will be written as,

Table 5.2 Floating leg of a vanilla swap in arrears

```
Receive Notional * (ir:eur3m on Now - 0m)* DcFr on Schedule;
```

If we want to make our swap callable, we add the following line of code, which defines the cashflows Y and the dates when they need to be paid.

Table 5.3 Swap callable at each payment date specified in the swap schedule

```
Long Callable on Schedule into (Receive 0 EUR on Schedule);
```

If we have the option to enter into a swap, i.e. if we want to have a trade with physical exercise, we can write the code below, where the non-intrinsic feature of the callability statement is explicit.

Table 5.4 Physically settled swaption

```
Notional = 100mm USD;
Schedule = From 2009/09/30 to 2019/06/30 every 6 months;
Date = 2014/09/30;
DcFr = DCF(now-6m, now, "ACT/365");
Swap = Receive Notional*(IR:USD6M on now-6m)*DcFr on Schedule;
Swap = Pay Notional * 3%* DcFr on Schedule;
// The settlement is physical
Long Callable on Date into Swap with Physical Settlement
   NonIntrinsic;
```

Consider now the example of a non-callable product, e.g. the barrier for which we have also shown the expression tree in Fig. 5.1. The PAL code could be written as follows,

Table 5.5 FX up and out option

```
Notional = 100mm EUR;
Strike = 1.0;
Barrier = 1.5;
Receive Notional * max(fx:gbpeur - Strike, 0.0) *
   (maximum(fx:gbpeur, 2009/10/20, 2010/10/20) < Barrier ?
   1.0 : 0.0) on 2010/10/20;
```

An example of a CDS on a counterparty characterised by a 'Credit Curve', is given in the table below. We use here two constructs, 'creditloss' and 'creditevents'. The first corresponds to the loss suffered by the underlying instrument and it is used to describe the protection leg of the CDS. The second indicates simply the event of default and therefore does not take into account recovery rate. It is used to define the payment leg of the CDS. More details are given in Chap. 6, where we describe how these quantities are taken from underlying simulated processes, and in Chap. 10, where we focus on credit derivatives.

We have mentioned that PAL allows also to use other typical programming language features, such as loops, matrix operations and some typical object-oriented construction. These features can significantly help the booking of complex

Table 5.6 CDS

```
start = 2004/07/20;
end = 2009/09/20;
Schedule = From (start+3m) to end every 3 months;
Notional = 100 mm EUR;
DcFr = DCF(now-3m, now, "ACT/ACT"); // Day Count Fraction
Receive Notional * creditloss(cr:"CreditCurve", Now - 3m, now)
    on Schedule;
Pay Notional * 0.0043 *
    (1.0-creditevents(cr:"CreditCurve", start, now))*DcFr
    on Schedule;
```

transactions, described usually in long termsheets. Below is the example of a trade using a 'for loop' to describe repetitive payments, and the definition of a new product, a swap, which can be used later within the program.

Table 5.7 For Loop example

```
Notional = 100mm USD;
DcFr1 = DCF(now-6m, now, "ACT/ACT");
DcFr2 = DCF(now-1Y, now, "ACT/ACT");
ScheduleRec = From 2009/01/01 to 2020/01/01 every 1y;
SchedulePay = From 2009/01/01 to 2020/01/01 every 6m;
Maturity=2020/01/01;
Receive Notional * 0.04 * DcFr2 on ScheduleRec;
Pay Notional * (((USD 10y)-(USD 2y))>0
    ? ((USD 6M)+0.007)*DcFr1 : 0) on SchedulePay;
DD=2005/04/22;
S=0;
For (i=1;i<60;i=i+1) {
    DD+=6m;
    S=S+(((USD 10y on DD)-(USD 2y on DD))>0
        ? ((USD 6M on DD)+0.007)*DcFr1 : 0);
}
Pay Notional * Max(0,0.3-S) on Maturity;
```

The code snippet above represents a complex interest-rate trade, where party A pays a fix rate every year, and party B pays Libor plus spread or zero every six months, depending on the difference between two points of the swap curve, the 10 years and the 2 years points. In addition, at maturity, a cumulative coupon is paid. This is computed within the 'for loop' depending again on the two points of the curve.

For very simple products, whose valuation is model-independent (such as for example standard interest-rate swaps), it is possible to use functions, which we call 'Instruments' (see below) to define an analytical pricing function. This gives the possibility of combining AMC with other pricing techniques. We will see at the end of this chapter how to use 'Instruments' to define more complicated products.

Table 5.8 Instrument example

```
Instrument Swap(Notional, Currency, Fix, FromDate, ToDate, Freq)
{
        DcFr = DCF(now-Freq, now, "ACT/ACT");
        Float = Libor(Currency, now - Freq, now - Freq, now);
        Schedule = From (FromDate + Freq) to ToDate every Freq;
        Receive Notional * Fix * DcFr Currency on Schedule;
        Pay Notional*(Float on now - Freq)*DcFr Currency
            on Schedule;
}
Buy Swap(100 mm, CHF, 3%, 2009/06/30, 2020/06/30, 6 m);
```

5.3 The Concept of Scenarios

We have mentioned on many occasions that the classical Monte Carlo framework used to compute counterparty exposure is via scenario generation and then pricing using analytical formulas or approximations. Sometimes these scenarios are generated in a centralised location and then sent to various engines that perform the pricing step.

AMC can also be considered as a pricing approximation, even if sophisticated and thus enabling the valuation of complex transactions. In this sense we could think of scenario generation and AMC valuation as two separated sub-systems. However, because we want simulation and pricing treatment to be generic, it is not possible to know beforehand which financial quantities will need to be extracted from the simulation. Therefore it is essential to be able to retrieve any financial quantity efficiently from the basic stochastic drivers.

5.4 The Concept of Super-Product

With the computational framework we have described and the definition of the PAL language, we can now make a step further, and define new types of products, whose payoffs, i.e. X and Y cashflows, depend on the price distribution of an already computed product. We call these products *Super-Products* to highlight the fact that these are products built with the results of the computation performed for another product.

5.4.1 An Example of Super-Products: The C-CDS

An example of a super-product is the contingent credit default swap (C-CDS). We will analyse this product in detail in Chap. 14, where we show how to compute credit valuation adjustments and how to hedge counterparty credit exposure. In this context it is sufficient to note that a C-CDS is an OTC derivative between two counterparties, A and B, say. Assume A has a derivative portfolio with a third counterparty C, and that it enters into a C-CDS with B. In case of default of C, under the C-CDS contract B will pay to A the positive value (as seen from A's perspective), V_t^+, of the portfolio. In other words a C-CDS corresponds to the protection leg of a CDS paying at each point in time the value of an underlying transaction. The construction of the C-CDS is performed as follows:

(i) first the price distribution V_t of the derivative portfolio is computed,
(ii) then defaults of counterparty C are simulated, and,
(iii) finally the cashflows of the C-CDS are created combining the default values with the price distribution.

The advantage of a generic computational framework, where only cashflows are relevant and valuation is performed via AMC are clear. We can effortlessly create a new product whose payoff depends on another product.

In Table 5.9 we show a PAL example of a C-CDS. We use several concepts described before: object-oriented features, the idea of 'Instrument' and the construct used to define CDSs. We can see in particular that the C-CDS corresponds, as mentioned above, to the protection leg of a CDS.

Table 5.9 CCDS

```
Instrument ccds(objInstrument)
{
    receive max(0, objInstrument(currentdate)) *
        creditevents(objInstrument.cpty,previousdate,currentdate))
        on objInstrument.schedule);
}
```

Chapter 6
Implementation

The previous chapter introduced a computational framework within which complicated payoffs can be specified and then simulated to obtain the price distributions required for credit exposure estimation. Trade specification is based on quantities we called *statistics*, which can be thought of as functions that return some financial quantity, given a simulated scenario. We will use these statistics later in Part III to specify various products.

This chapter is dedicated to a more detailed analysis of various statistics. We describe their implementation, the practical issues that arise, and the solutions we adopted. Since simulation is at the heart of our framework, we describe also various Monte Carlo schemes for simulating SDEs. We end the chapter by analysing the different types of errors introduced in the various steps of the modelling.

6.1 Spot and Forward Statistics

The first type of statistics we consider are those that extract spot and forward values directly from simulated scenarios. These statistics are relatively simple in the sense of being deterministic functions of the simulated Brownian Motion processes. Typical examples are values of stock and foreign exchange rates, bond prices, and Libor rates. Most of these statistics can have a common signature and can be differentiated by the type of underlying they are applied to. Thus, we define a generic statistic (called INDEX), which has as argument the type of underlying (e.g. EQ, FX, or IR), and a symbol identifying a specific instance of the underlying (e.g. the IBM stock, the USDGBP exchange rate, the USD3Y three years swap rate). Having defined the underlying type and its instance, the observation date is expressed as a *lag* relative to the payoff date. Examples are given in Table 6.1.

The generic way we have implemented the INDEX statistic is as follows. Different underlying types will correspond to different martingales, each of which are deterministic functions of simulated Brownian Motion paths. All that the INDEX statistic needs to do is to look up the correct martingale for each underlying type that is asked for.

G. Cesari et al., *Modelling, Pricing, and Hedging Counterparty Credit Exposure,*
Springer Finance, DOI 10.1007/978-3-642-04454-0_6,
© Springer-Verlag Berlin Heidelberg 2009

Table 6.1 At payoff date, return the value of index SYMBOL, of type TYPE, observed at the payoff date minus lag LAG. Examples:
(i) The 3-year USD swap rate is defined as USD3Y, and its type, "interest rate", as IR. In a 3Y USD CMS swap contract, the swap rate is called as INDEX(IR, USD3Y, 0M).
(ii) To define a vanilla interest rate swap, paying quarterly, the 3-month libor rate fixed 3 months ago is accessed as INDEX(IR, USD3M, 3M).
(iii) The IBM stock is called as INDEX(EQ, IBM, 0M).
(iv) USDGBP FX rate is called as INDEX(FX, USDGBP, 0M).
(v) USDGBP 1 year forward is called as INDEX(FX, USDGBP, -1Y)

	Date	Payoff
Spot and forward	T_i	INDEX(TYPE, SYMBOL, LAG)

The reader may have noticed one practical problem that needs to be dealt with, and will affect computation of all statistics. When simulating the basic martingales for different asset classes, the simulation will need to happen on a finite number of prespecified time-points $t \in \{t_0, t_1, \ldots, t_n\}$. Of course the t_j cannot be chosen *a priori* to incorporate all payoff dates of the portfolio that needs to be computed, first because there might be too many such dates, and second because to ensure scenario consistency one needs to be able to compute price distributions of new portfolios starting from the same basic Brownian Paths. We solve this problem by an interpolation method which we describe in a later section in this chapter.

6.1.1 Libor Rates and Bond Prices

Libor rates and bond prices occur frequently and play an important role in a large proportion of portfolios. For this reason we have defined specialised statistics to aid in creating payoffs that depend on these quantities.

Libor Rates are a specific case of an INDEX in our computational framework, but we have nevertheless defined a specialised statistic that returns simulated Libor rates. Recall that if $t < [T_1, T_2]$, the Libor rate $L_{t,[T_1,T_2]}$ observed at time t for the period $[T_1, T_2]$, is a simple function of bond prices, that is,

$$L_{t,[T_1,T_2]} = \frac{1}{T_2 - T_1} \left(\frac{D_{t,T_1}}{D_{t,T_2}} - 1 \right) = \frac{1}{T_2 - T_1} \left(\frac{D_{0,T_1}}{D_{0,T_2}} \frac{M_{t,T_1}}{M_{t,T_2}} - 1 \right), \tag{6.1}$$

where we have used the representation of bond prices

$$D(t, T) = \frac{D(0, T)}{D(0, t)} \frac{M(t, T)}{M(t, t)} \tag{6.2}$$

in terms of the basic martingales M. Table 6.2 shows the syntax for both Libor and bond price statistics.

Table 6.2 At payoff date, return the Libor rate for a specified currency and tenor observed at payoff date minus lag. Example: the 6-months EUR Libor rate fixed in arrears is called `LIBOR(EUR, 6M, 0M)`. Similarly, `ZEROBOND(EUR, 10Y, 1Y)` returns the price of a 10-year bond as observed 1 year before the current payoff date

	Date	Payoff
Libor	T_i	`LIBOR(CURRENCY, TENOR, LAG)`
ZeroBond	T_i	`ZEROBOND(CURRENCY, MATURITY, LAG)`

6.1.2 Annuity

An annuity pays a unit coupon at regular points in time. In other words, it pays the difference between a coupon bond and a zero bond. The implementation of this statistic is similar to the previous one.

$$A_{t,T_1,\dots,T_n} = \sum_{i=1}^{n} (T_i - T_{i-1}) D_{t,T_i} = \sum_{i=1}^{n} (T_i - T_{i-1}) D_{0,T_i} \frac{M_{t,T_i}}{M_{t,t}}. \qquad (6.3)$$

Table 6.3 shows the usage of the `ANNUITY` function.

Table 6.3 At payoff date, return the annuity of a swap with specified tenor and payment frequency. Example: the current annuity of a 3-years USD swap with monthly payments is specified as `ANNUITY(USD, 3Y, 12, 0M)`

	Date	Payoff
Annuity	T_i	`ANNUITY(CURRENCY, MATURITY, TIMESPERYEAR, LAG)`

6.1.3 Swap Rate

The swap rate is the coupon that the fixed-rate leg of a swap would have to pay for the swap to have zero net present value. Said in another way, it is the coupon that an annuity would need to pay in order to have value equal to a floating-rate leg. Thus, the `SWAPRATE` function can be evaluated as

$$s_{t,T_1,\dots,T_n} = \frac{D_{t,T_0} - D_{t,T_n}}{A_{t,T_1,\dots,T_n}} = \frac{D_{0,T_0} M_{t,T_0} - D_{0,T_n} M_{t,T_n}}{A_{t,T_1,\dots,T_n} M_{t,t}}. \qquad (6.4)$$

Again, the right side above involves only quantities that have already been considered. A specific example of usage of `SWAPRATE` is shown in Table 6.4.

Table 6.4 At payoff date, return the par rate of a swap with specified tenor and payment frequency. Example: the par rate of a 3-year USD swap with monthly payments is called SWAPRATE(USD, 3Y, 12, 0M)

	Date	Payoff
Swap Rate	T_i	SWAPRATE(CURRENCY, TENOR, TIMESPERYEAR, LAG)

6.2 Path Dependent Statistics

We now turn to statistics whose evaluation cannot be done in simple deterministic fashion from simulated Brownian Motion paths. Examples of such statistics are the maximum attained by a process, the average value of a process, and the time spent by a process within a given range. In principle, if one could simulate the basic Brownian Motions on a fine-enough grid, such statistics would be a sampling exercise. However, since the size of portfolios and time horizons involved do not allow the luxury of arbitrarily small time steps, we need to find estimators of the above quantities that remain accurate when the simulation time step is relatively large.

6.2.1 Extremum

Extremum is the maximum or minimum value reached on a given scenario path between two observable dates. This statistic is typically used to describe barrier features. The parameters we need for its implementation are its type and symbol, and some time parameters to define where to perform the observation.

Table 6.5 At payoff date, return either the maximum (ISMAX = true) or the minimum (ISMAX = false) observed value of an index over an observation period starting at $Tstart = Tpay - lag - tenor$, and finishing at $Tend = Tpay - lag$

	Date	Payoff
Extremum	T_i	EXTREMUM(TYPE, SYMBOL, LAG, TENOR, ISMAX)

The implementation of EXTREMUM can be performed using some properties of Brownian motions. Consider an $\mathbb{N}$-Brownian motion W for which we need to estimate the maximum and minimum values on an interval $[t, T]$, and suppose that we know the values, $W_t = a$ and $W_T = b$. The idea is to consider maxima and minima as random variables, defined by,

$$W_{max} = \max_{u \in [t,T]} W_u \quad \text{and} \quad W_{min} = \min_{u \in [t,T]} W_u. \quad (6.5)$$

We can write (see the Technical Note 6.4.4),

$$\mathbb{E}(W_{max} \mid W_t = a, W_T = b) = \frac{a+b}{2} + \sqrt{\frac{T-t}{2}}\left[\sqrt{K} + e^K \sqrt{\pi}\Phi(-\sqrt{2K})\right] \quad (6.6)$$

$$\mathbb{E}\left(W_{min} \mid W_t = a, W_T = b\right) = \frac{a+b}{2} - \sqrt{\frac{T-t}{2}}\left[\sqrt{K} + e^K \sqrt{\pi}\Phi(-\sqrt{2K})\right], \quad (6.7)$$

where

$$K = \frac{(b-a)^2}{2(T-t)}. \quad (6.8)$$

Once we have simulated the extremum of the underlying stochastic driver, as long as the quantity X (e.g. stock price, libor rate...) can be expressed as monotonic function f of a single Brownian motion, we can derive the maximum (resp. minimum) of X as being either $f(W_{max})$ (resp. $f(W_{min})$) if $f' > 0$ or $f(W_{min})$ (resp. $f(W_{max})$) if $f' < 0$.[1]

6.2.2 Average

Knowing X_{max} and X_{min} on any interval allows also to compute quantities that are path dependent, such as average values, or days within a range. A simple way of doing so for any interval $[T_1, T_2]$ is

- From X_{T_1} and X_{T_2} obtain X_{max} and X_{min}.
- Interpolate the path $(T_1 \to T_2)$ by imposing for example that max and min occur at $(T_2 - T_1)/4$ and $3(T_2 - T_1)/4$,

$$X\left(\frac{T_2 - T_1}{4}\right) = X_{max}, \quad \text{and} \quad X\left(\frac{3(T_2 - T_1)}{4}\right) = X_{min}. \quad (6.9)$$

Polynomial interpolation can now be used to obtain the path of X on $[T_1, T_2]$ from the four values given.

Table 6.6 At payoff date, return the average value of any index over an observation period starting at $Tstart = Tpay - lag - tenor$, and finishing at $Tend = Tpay - lag$

	Date	Payoff
Average	T_i	AVERAGE(TYPE, SYMBOL, LAG, TENOR)

Asian options require to simulate the average value $\Pi_{Average}$ of X on a time interval $[T_1, T_2]$:

$$\Pi_{Average} = \frac{1}{T_2 - T_1}\int_{T_1}^{T_2} X_u du. \quad (6.10)$$

---

[1]Note that this approximation gives a bias due to Jensen's inequality. If needed, an approximation for $\mathbb{E}(f_{max})$ on the Brownian bridge could be implemented. This, however, becomes model dependent.

6.2.3 In Range Fraction

Range accrual products usually pay an exotic coupon, which will be the product of a fixed coupon c with the proportion of days a given underlying has remained within a range $[K_{lower}, K_{upper}]$.

$$\Pi_{InRange} = \frac{c}{T_2 - T_1} \int_{T_1}^{T_2} \mathbb{1}_{X_u \in [K_{lower}, K_{upper}]} du, \quad \in [0, c]. \quad (6.11)$$

Table 6.7 At payoff date, return n/N, where n is the number of days where the index is in the range $[Lower\,Range, Upper\,Range]$, and N is the total number of days in the observation period. The observation period starts at $T\,start = T\,pay - lag - tenor$, and finishes at $T\,end = T\,pay - lag$

	Date	Payoff
Average	T_i	INRANGEFRACTION (TYPE, SYMBOL, LAG, TENOR, LOWERRANGE, UPPERRANGE)

6.2.4 Credit Loss

To compute credit derivatives we need to be able to compute credit losses occurring in a given time interval. They are defined as follows,

$$Loss_{T_i, T_{i+1}} = \frac{\sum_{j=1}^{n} N_j (1 - R_j) \mathbb{1}_{T_i < \tau_j \le T_{i+1}}}{\sum_{j=1}^{n} N_j}, \quad (6.12)$$

where N_i is the notional and R_i the recovery rate of name i. To describe some products it is also useful to define the proportion of defaulted notional,

$$CreditEvent_{T_i, T_{i+1}} = \frac{\sum_{j=1}^{n} N_j \mathbb{1}_{T_i < \tau_j \le T_{i+1}}}{\sum_{j=1}^{n} N_j}. \quad (6.13)$$

Table 6.8 At payoff date, return the loss occurring between pay-off date minus lag and tenor

	Date	Payoff
Credit Loss	T_i	CREDITLOSS (REF, LAG, TENOR)

6.3 Monte Carlo Stepping

At the core of our work is Monte Carlo simulation. There is a vast literature on SDE integration (see for example the books by Glasserman [50], Kloeden & Platen [70], Milstein [82], Øksendal [83], and Protter [86]). We will give here only an overview, highlighting some practical aspects related to the specific equation we deal with.

If the SDEs of the underlying process have a closed form solution, it is possible to implement the Monte Carlo stepping without regards to convergence. However, if the form of the process is more complicated, numerical schemes will have to be used to provide the stepping algorithm.

Consider a stochastic process X driven by the following stochastic differential equation,

$$dX_t = a(X_t, t)dt + b(X_t, t)dW_t. \tag{6.14}$$

To simulate it, we can break the simulation schedule into a finite set of dates $\{t_i\}$ such that $\Delta_i := t_{i+1} - t_i$. The most straightforward approximation to (6.14) is to use the Euler scheme

$$X_{i+1} = X_i + a(X_i, t_i)\Delta_i + b(X_i, t_i)\sqrt{\Delta_i}z, \quad z \sim N(0, 1). \tag{6.15}$$

This scheme is inaccurate if simulations with large Δ_i, are required. In these cases we can use integration schemes of higher order in Δ_i. Consider the solution to (6.14) between t_i and t_{i+1}. It is given by

$$X_{t_{i+1}} = X_{t_i} + \int_{t_i}^{t_{i+1}} a(X_u, u)du + \int_{t_i}^{t_{i+1}} b(X_u, u)dW_u. \tag{6.16}$$

To compute numerically this equation, we need to approximate the two integrals on the right hand side. This can be achieved by using Itô's lemma. Indeed, applying Itô's formula to any C^2 function $f(X, t)$, we get,

$$f(X_u, u) = f(X_{t_i}, t_i) + \int_{t_i}^u \frac{\partial f}{\partial X}(X_s, s)a(X_s, s)ds + \int_{t_i}^u \frac{\partial f}{\partial X}(X_s, s)b(X_s, s)dW_s$$

$$+ \frac{1}{2}\int_{t_i}^u \frac{\partial^2 f}{\partial X^2}(X_s, s)b^2(X_s, s)ds + \int_{t_i}^u \frac{\partial f}{\partial t}(X_s, s)ds. \tag{6.17}$$

We can apply (6.17) to a and b in (6.16), limiting the expansion to stochastic integrals having a variance of order 4 in time. We obtain then,

$$X_{t_{i+1}} = X_{t_i} + a\Delta_i + b\left(W_{t_{i+1}} - W_{t_i}\right) \tag{6.18}$$

$$+ b'bI_{(1,1)} \tag{6.19}$$

$$+ \left[b'a + \frac{b''b^2}{2} + \dot{b}\right]I_{(0,1)} + a'bI_{(1,0)} + \left[b''b^2 + b\left(b'\right)^2\right]I_{(1,1,1)} \tag{6.20}$$

$$+\begin{cases}\left(a'a+\frac{a''b^2}{2}+\dot{a}\right)I_{(0,0)}+\left(b'''b^2+4bb'b''+(b')^3\right)I_{(1,1,1,1)}\\ +\left[bb''a+ba'b'+\frac{b}{2}\left(b'''b^2+2bb'b''\right)+\dot{b}'b\right]I_{(1,0,1)}\\ +\left(a''b^2+a'b'b\right)I_{(1,1,0)}\\ +\left[ab''b+ab'b'+\frac{1}{2}\left(b'''b^3+3b''b'b^2\right)\right]I_{(0,1,1)},\end{cases}\quad(6.21)$$

where $'$ denotes a derivative with respect to X and $\dot{}$ denotes a derivative with respect to t; all functions are evaluated at $t = t_i$.

Define the multiple integrals $I_{(\delta_i)}$ to have all integrands equal to one and integrator being either W if $\delta_i = 1$, or t otherwise. For example,

$$I_{(1,1,0)} = \int_{t_i}^{t_{i+1}} \int_{t_i}^u \int_{t_i}^s dW_v dW_s du.$$

The first line (6.18) is the usual Euler scheme. We can break the remaining three lines into three terms in orders of Δ:

$$(6.19) \sim o(\Delta)$$

$$(6.20) \sim o\left(\Delta^{3/2}\right)$$

$$(6.21) \sim o\left(\Delta^2\right).$$

We can expand these integrals (see Technical Note 6.4.1), to obtain the following results.

(i) *Euler Scheme.* The general Euler scheme translates (6.14) into:

$$X_{t_{i+1}} = X_{t_i} + a_{t_i}\Delta + b_{t_i}\sqrt{\Delta}z. \qquad (6.22)$$

This scheme has a convergence order of $O(\Delta)$. Recall that in (6.19), we have unveiled another term whose convergence is $O(\Delta)$. Therefore, the Euler scheme is incomplete, and should be expanded to,

(ii) *Milstein 1 Scheme.* This scheme translates (6.14) into,

$$X_{t_{i+1}} = X_{t_i} + a_{t_i}\Delta + b_{t_i}\sqrt{\Delta}z + \frac{1}{2}b'_{t_i}b_{t_i}\Delta\left(z^2 - 1\right). \qquad (6.23)$$

This scheme is the true SDE scheme of order one in Δ for (6.14). Higher orders of convergence are described in the following schemes.

(iii) *Milstein 3/2 Scheme.* This scheme translates (6.14) into:

$$X_{t_{i+1}} = X_{t_i} + a_{t_i}\Delta + b_{t_i}\sqrt{\Delta}z + \frac{1}{2}b'_{t_i}b_{t_i}\Delta(z^2 - 1)$$

$$+\begin{cases}a'_{t_i}b_{t_i}\Delta^{3/2}(z - \frac{v}{\sqrt{3}})\\ +[b'_{t_i}a_{t_i} + \frac{b''_{t_i}b^2_{t_i}}{2} + \dot{b}_{t_i}]\Delta^{3/2}(z + \frac{v}{\sqrt{3}})\\ +[b''_{t_i}b^2_{t_i} + b_{t_i}(b'_{t_i})^2]\frac{\Delta^{3/2}z}{6}(z^2 - 3),\end{cases}\quad(6.24)$$

where z and v are two iid standard normal distributions. This scheme converges to an order $3/2$ in Δ.

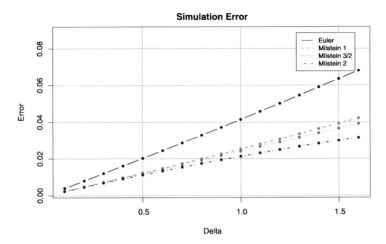

Fig. 6.1 Error f_{err} for each order in the SDE approximation; $\sigma = 30\%$ and $\mu = 5\%$

To assess how choosing an appropriate scheme can increase the accuracy of our results, consider the following simple SDE:

$$dS_t = \mu S_t dt + \sigma S_t dW_t,\qquad(6.25)$$

for which we have $a(S_t, t) = \mu S_t$ and $b(S_t, t) = \sigma S_t$. The true solution to this equation is,

$$S_t = S_0 e^{(\mu - \frac{\sigma^2}{2})t + \sigma W_t}.\qquad(6.26)$$

We define an error function f_{err} by

$$f_{err}(\Delta) = \sqrt{\mathbb{E}\left[\left(S_0 e^{(\mu - \frac{\sigma^2}{2})\Delta + \sigma\sqrt{\Delta}z} - \hat{S}(\Delta)\right)^2\right]},\quad \Delta \geq 0,\qquad(6.27)$$

where $\hat{S}$ is the approximated solution to (6.14) evaluated using one of the four SDE schemes we have described and Fig. 6.1 compares the values of f_{err} for the four schemes. The simulated value for S should always be strictly positive. However, depending on which scheme one chooses, this may not be satisfied. Figure 6.2 shows how in this simple case, choosing Euler leads to non-zero probabilities of having $S_t < 0$.

In practice we have chosen the Milstein 1 scheme to perform the integration of our SDEs. From Fig. 6.1 we can see that, in our simple example, any Milstein scheme outperforms significantly the classical Euler scheme.

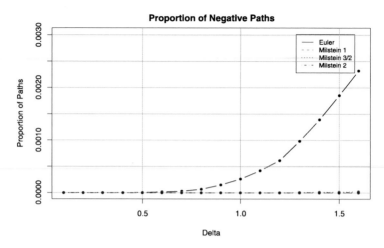

Fig. 6.2 Proportion of paths being below zero for each order in the SDE approximation; $\sigma = 30\%$ and $\mu = 5\%$ (see 6.4.2 for details on Milstein 2)

6.4 Technical Notes

In the sections below we give technical details of some results we have used previously.

6.4.1 SDE Integration Schemes

We write the integrals I in (6.18) in terms of Δ and $\Delta W = W_{t_{i+1}} - W_{t_i} = \sqrt{\Delta}z$, with z being a standard normal distribution. After some calculus we obtain:

$$I_{(0,0)} = \frac{\Delta^2}{2} \tag{6.28}$$

$$I_{(0,1)} + I_{(1,0)} = \Delta^{3/2}z \tag{6.29}$$

$$I_{(1,1)} = \frac{\Delta}{2}\left(z^2 - 1\right) \tag{6.30}$$

$$I_{(1,1,1)} = \frac{\Delta^{3/2}z}{6}\left(z^2 - 3\right) \tag{6.31}$$

$$I_{(1,1,0)} + I_{(1,0,1)} + I_{(0,1,1)} = \frac{\Delta^2}{2}\left(z^2 - 1\right) \tag{6.32}$$

$$I_{(1,1,1,1)} = \frac{\Delta^2}{4}\left(\frac{z^4}{6} - z^2 + \frac{1}{2}\right). \tag{6.33}$$

Introducing a second standard normal random variable v, independent from z, we can rewrite the integrals as:

$$I_{(1,0)} = \frac{\Delta^{3/2}}{2} \left(z + \frac{v}{\sqrt{3}} \right) \tag{6.34}$$

$$I_{(1,0)} = \frac{\Delta^{3/2}}{2} \left(z - \frac{v}{\sqrt{3}} \right). \tag{6.35}$$

Introducing two more standard normal random variable ω and $\tilde{\omega}$, independent from each other, we can rewrite the integrals as:

$$I_{(1,1,0)} = \frac{\Delta^2}{4} \left(\omega^2 - 1 \right) \tag{6.36}$$

$$I_{(1,0,1)} = \frac{\Delta^2}{4} \left[z^2 - \frac{\omega^2 - 1}{2} - \frac{8}{\sqrt{3}} \left(\tilde{\omega}^2 - 1 \right) \right] \tag{6.37}$$

$$I_{(0,1,1)} = \frac{\Delta^2}{4} \left[z^2 - \frac{\omega^2 - 1}{2} + \frac{8}{\sqrt{3}} \left(\tilde{\omega}^2 - 1 \right) \right]. \tag{6.38}$$

6.4.2 Milstein 2 Scheme

We can further expand the integrals seen in Sect. 6.3 to obtain the so-called Milstein 2 scheme.

$$X_{t_{i+1}} = X_{t_i} + a_{t_i} \Delta + b_{t_i} \sqrt{\Delta} z + \frac{1}{2} b'_{t_i} b_{t_i} \Delta(z^2 - 1)$$

$$+ \begin{cases} a'_{t_i} b_{t_i} \Delta^{3/2} (z - \frac{v}{\sqrt{3}}) \\ \quad + [b'_{t_i} a_{t_i} + \frac{b''_{t_i} b^2_{t_i}}{2} + \dot{b}_{t_i}] \Delta^{3/2} (z + \frac{v}{\sqrt{3}}) \\ \quad + [b''_{t_i} b^2_{t_i} + b_{t_i} (b'_{t_i})^2] \frac{\Delta^{3/2} z}{6} (z^2 - 3) \end{cases}$$

$$+ \begin{cases} (a'a + \frac{a''b^2}{2} + \dot{a}) \frac{\Delta^2}{2} \\ \quad + (b'''b^2 + 4bb'b'' + (b')^3) \frac{\Delta^2}{4} (\frac{z^4}{6} - z^2 + \frac{1}{2}) \\ \quad + [bb''a + ba'b' + \frac{b}{2}(b'''b^2 + 2bb'b'') + \dot{b}'b] \\ \quad \times \frac{\Delta^2}{4} [z^2 - \frac{\omega^2 - 1}{2} - \frac{8}{\sqrt{3}}(\tilde{\omega}^2 - 1)] \\ \quad + (a''b^2 + a'b'b) \frac{\Delta^2}{4} (\omega^2 - 1) \\ \quad + [ab''b + ab'b' + \frac{1}{2}(b'''b^3 + 3b''b'b^2)] \\ \quad \times \frac{\Delta^2}{4} [z^2 - \frac{\omega^2 - 1}{2} + \frac{8}{\sqrt{3}}(\tilde{\omega}^2 - 1)], \end{cases} \tag{6.39}$$

with ω and $\tilde{\omega}$ being two iid standard normal distributions.

6.4.3 Martingale Interpolation

The need to evaluate given statistics or functions at any payoff time specified by a trade representation is faced by the restriction that Brownian Motion paths can only be simulated at finite points on a chosen grid. Moreover, computational tractability puts a lower bound on the simulation time step that can be chosen.

In view of the above, we need an interpolation scheme that makes available a 'simulated' value of a process (in particular, our basic martingale processes) at any time t, given simulated values on grid $\mathbb{T} = \{t_j, j = 0, 1, \ldots, n\}$ that does not include t.

In detail, consider the positive martingale process $M_{t,T}$ satisfying

$$dM_{t,T} = M_{t,T} \sigma_{u,T} dW_t, \tag{6.40}$$

where W is Brownian Motion. Now choose a time $t \in (t_j, t_{j+1})$, and assume that the Brownian Motion W has been simulated on the grid $\mathbb{T}$. In particular, since

$$\log M_{t,T} = \log M_{0,T} + \int_0^t \sigma_{u,T} dW_u - \frac{1}{2} \Sigma_{t,T}^2 t, \tag{6.41}$$

we can then assume that simulated values of $M_{t_j,T}$ and $M_{t_{j+1},T}$ are available. The idea is now to define the unknown value $\hat{M}_{t,T}$ in terms of the closest simulated martingale values $M_{t_j,T}$ and $M_{t_{j+1},T}$ in such a way as to respect the martingale property of M. First, note that

$$M_{t,T} = M_{t_j,T} \exp\left(\int_{t_j}^t \sigma_{u,T} dW_u - \frac{1}{2} \Sigma_{t_j,t,T}^2 (t - t_j) \right), \tag{6.42}$$

where

$$\Sigma_{s,t,T}^2 (t - s) := \Sigma_{t,T}^2 t - \Sigma_{s,T}^2 s \tag{6.43}$$

is seen to be the variance of the increment of $\log M$ over the interval $[s, t]$. Because the values of W are not available in $(t_j, t]$, we propose to define a value $\hat{M}_{t,T}$ through

$$\hat{M}_{t,T} := M_{t_j,T} \exp\left(\Sigma_{t_j,t,T} \left(a W_{t_j} + b W_{t_{j+1}} \right) - \frac{1}{2} \Sigma_{t_j,t,T}^2 (t - t_j) \right),$$

in terms of *known* W_{t_j} and $W_{t_{j+1}}$. The criterion for choosing the weights a and b is the martingale property for M at t, namely that

$$\mathbb{E}[\hat{M}_{t,T} \mid \mathscr{F}_{t_j}] = M_{t_j,T}. \tag{6.44}$$

By taking expectations on both sides of the previous equation, and writing $W_{t_{j+1}} \equiv W_{t_j} + (W_{t_{j+1}} - W_{t_j})$ as the sum of two independent terms, we get that

$$\mathbb{E}[\hat{M}_{t,T} \mid \mathscr{F}_{t_j}] = M_{t_j,T} \exp\left(\Sigma_{t_j,t,T}(a + b) W_{t_j} - \frac{1}{2} \Sigma_{t_j,t,T}^2 (t - t_j) \right)$$

$$\times \mathbb{E}\left[\exp\left(b\Sigma_{t_j,t,T}(W_{t_{j+1}} - W_{t_j})\right) \mid \mathscr{F}_{t_j}\right]$$

$$= M_{t_j,T}\exp\left(\Sigma_{t_j,t,T}(a+b)W_{t_j} - \frac{1}{2}\Sigma_{t_j,t,T}^2(t - t_j)\right)$$

$$\times \exp\left(\frac{1}{2}b^2\Sigma_{t_j,t,T}^2(t_{j+1} - t_j)\right). \tag{6.45}$$

The only way that the above coincides with $M_{t_j,T}$ is by having

$$a = -b$$

$$b^2 = \frac{t - t_j}{t_{j+1} - t_j}. \tag{6.46}$$

6.4.4 Distribution of Maxima and Minima

We discuss here the estimation of the maxima and minima attained by a Brownian path over a chosen time interval, bearing in mind that the Brownian path simulation is available only at a discrete (possibly widely-spaced) set of time points $\mathbb{T} = \{t_j, j = 0, 1, \ldots, n\}$. A natural way to do this is to consider the law of the maximum and minimum attained by the Brownian Bridge between any two consecutive time points t_k, t_{k+1}.

Without loss of generality, let X be standard Brownian Motion, and let $B \equiv (B_u)_{0 \leq u \leq T}$ be X conditioned to be at $b > 0$ at time T. The laws of the processes $\bar{B}$ and $\underline{B}$, defined by

$$\bar{B}_u = \max\{B_s, 0 \leq s \leq u\}, \quad u \leq T, \tag{6.47}$$

$$\underline{B}_u = \min\{B_s, 0 \leq s \leq u\}, \quad u \leq T, \tag{6.48}$$

are easily derived using the reflection principle (see, for example, [68]) and shown to be distributions of Rayleigh type. Indeed, we have the laws

$$\mathbb{N}\left(\bar{B}_T \leq H\right) = 1 - \exp\left(-2H(H - b)/T\right), \quad H \geq b \tag{6.49}$$

$$\mathbb{N}\left(\underline{B}_T \leq -H\right) = \exp\left(-2H(H + b)/T\right), \quad H \geq 0. \tag{6.50}$$

Inverting the marginal laws of $\bar{B}_T$ and $\underline{B}_T$, we can write, in terms of a uniform variate U on the unit interval

$$\bar{B}_T \sim \frac{1}{2}b + \frac{1}{2}\sqrt{b^2 - 2T\ln U}, \tag{6.51}$$

$$\underline{B}_T \sim \frac{1}{2}b - \frac{1}{2}\sqrt{b^2 - 2T\ln U}. \tag{6.52}$$

Of course, the above characterizes only the *marginal* laws of the maximum and minimum of B—we say nothing about the joint law.

More generally, if B is a Brownian bridge on $[s, t]$ with $B_t = x$ and $B_T = y$, we apply the arguments above to

$$\{B_u - x, s \le u \le t\}$$

if $y \ge x$, or, if $y < x$, to

$$\{x - B_u, s \le u \le t\},$$

which in the respective cases are Brownian Bridges starting at 0 and terminating at the positive value $|y - x|$ as assumed in our derivation above.

As mentioned in the description of the EXTREMUM function, certain trade payoffs are related to maxima and minima of functions of Brownian paths. The values of such payoffs can be approximated once we use (6.51) to write down expressions for the means of $\bar{B}$ and $\underline{B}$. To do so, we need to evaluate integrals of the form

$$I_K(\alpha) = \int_0^1 (K - \ln u)^\alpha \, du = \int_K^\infty e^{K-x} x^\alpha dx$$
$$= e^K \Gamma(\alpha + 1, K), \tag{6.53}$$

where our interest is in setting $K = b^2/(2T)$ and $\alpha = \frac{1}{2}$, and where

$$\Gamma(\alpha + 1, K) = \int_K^\infty x^\alpha e^{-x} dx \tag{6.54}$$

is the upper incomplete gamma function.

Using standard relations for gamma functions of different orders α, we get that

$$I_K(\alpha) = e^K \left(\alpha \Gamma(\alpha, K) + K^\alpha e^{-K} \right)$$
$$= e^K \alpha \Gamma(\alpha, K) + K^\alpha. \tag{6.55}$$

For the case that interests us here,

$$\Gamma\left(\frac{1}{2}, K\right) = 2 \int_{\sqrt{K}}^\infty e^{-t^2} dt = 2\sqrt{\pi} \Phi(-\sqrt{2K}), \tag{6.56}$$

using which we can finally write

$$\mathbb{E}\left(\bar{B}_T\right) = \frac{1}{2}b + \sqrt{\frac{T}{2}} \left[\sqrt{K} + e^K \sqrt{\pi} \Phi(-\sqrt{2K})\right], \tag{6.57}$$

$$\mathbb{E}\left(\underline{B}_T\right) = \frac{1}{2}b - \sqrt{\frac{T}{2}} \left[\sqrt{K} + e^K \sqrt{\pi} \Phi(-\sqrt{2K})\right]. \tag{6.58}$$

6.5 Error Analysis

Our modelling and AMC pricing methodology provides us with estimates of future price distributions. The main sources of error present in our modelling/pricing approach are summarised below.

(i) *Choice of model.* The models we have suggested are relatively simple, especially compared to the standard models used in the exotic interest rate market. Our choices have been mainly driven by the fact that models have to be applicable to different product types, in scenario-consistent fashion. It is natural to question, however, if the valuation that results from our model can be reconciled with prices and risk sensitivities that are used for actual hedging, i.e. if the model replicates market prices of hedging instruments.

(ii) *AMC error.* In the valuation framework, described in Chap. 4, price distributions are computed via AMC, which of course yields only a numerical estimate of the true price. For vanilla products, where closed-form valuation is possible, what is the discrepancy between the analytical value and that resulting form simulation? In the case of exotic products, what is the model risk introduced by this algorithm? In particular the AMC algorithm depends on the choice of observables. How robust is the algorithm through which the observable choice is made for each trade computation?

(iii) *Numerical error.* In addition to the error introduced because of choosing AMC as valuation algorithm, what is the numerical error introduced by the Monte Carlo simulation?

(iv) *Approximations.* In Chap. 2, where we described the modelling framework, we emphasised the possibility of assuming independence between interest rates and other asset classes. This allowed for a simplified modular framework in building models. As we also pointed out, the price of this simplicity is that arbitrage opportunities are introduced, for example in the case of modelling of FX rates. How severe is the actual numerical impact of this defect in the modelling? We have also mentioned that we can implement arbitrage free models at the cost of more complicated calibration. Is it worthwhile to perform this extra step?

Each of these questions deserve a very thorough analysis. In the following sections we will only briefly discuss some of these questions using specific examples, investigating the accuracy of our pricing and simulation.

6.5.1 Choice of Model: Scenario and Exposure Analysis

In general the price of derivatives is model dependent. Only simple vanilla products, such as for example vanilla interest-rate swaps, can be valued directly using market data. In other words, prices are subject to, what is called in the industry, *model risk*. When computing price distributions we have not only to choose a model for valuation, but also a model for generating scenarios.

To estimate pricing model risk we have compared our valuation with several other models. Without entering into details, we have found that the simple one-factor model we have defined in Chap. 3 gives in general satisfactory results, within the range of other benchmark models, such as HJM or BGM. Only some categories of products, which depend on different points of the yield curve, produce results that are not completely satisfactory. Examples of such products are steepeners, which we will describe in Chap. 8.

Credit derivatives, in particular CDO tranches offer similar challenges, as often prices are obtained by calibrating correlation of individual tranches on market observables. On the other hand a counterparty exposure system requires a model that can be used for all tranches simultaneously.

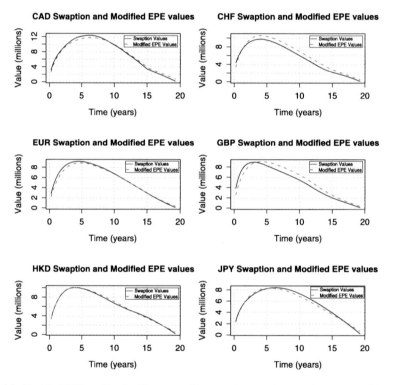

Fig. 6.3 Modified EPE profiles for 20-year vanilla swaps in several currencies, compared to prices of swaptions. The solid line shows swaption values, while the dotted one represents the modified EPE for the swap (computation performed April 09)

The problem of estimating potential errors in price distributions and their statistical measures (e.g. PFE or EPE) is more complicated, as in general there are no models available in the industry that can be used as standard benchmarks and, more importantly, there are no products trading these quantities. It is possible, however, to use market information to assess exposure profiles, when computing expected positive exposure. The value of EPE at time t can be seen in fact as the value of an option to enter at time t in the underlying product. Thus if option prices are available, it is possible to assess the quality of the model. We will see in Chaps. 12 and 14 that it is convenient to define a modified value of EPE to take into account this pricing feature. In Fig. 6.3 we compare the modified EPE profile of swaps, with the value of corresponding swaptions for different currencies.

To further analyse if the generated price distributions are realistic we can consider the underlying generated scenario.

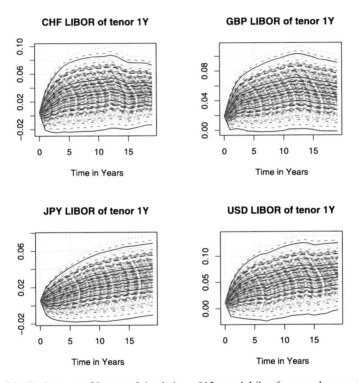

Fig. 6.4 Distributions, over 20 years of simulation, of 12-month Libor for several currencies. Each plot line shows a different percentile of the distribution, from 1% to 99% (computation performed April 09)

As a qualitative diagnostic for our modelling, we show graphically the distributions obtained for simulated Libor rates, swap rates, and foreign exchange rates, for several currencies. Due to the low-rates environment prevailing at the time of writing, for all currencies we see a small number of scenarios with negative Libor and swap rates. The FX rate for JPY shows a heavy skew, corresponding to what can also be observed from the data used to calibrate the model.

For comparison, Table 6.9 shows the historical extrema attained by Libor rates in the 7 years prior to the time of writing. For all currencies, the historical Libor values observed lie well within the range of values simulated in our framework.

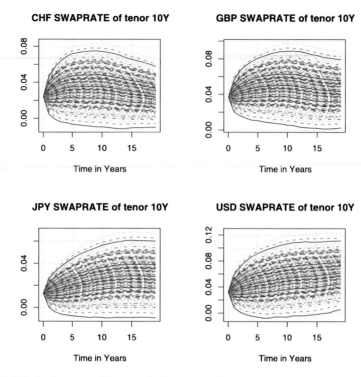

Fig. 6.5 Distributions, over 20 years of simulation, of the 10-year swaprate for several currencies. Each plot line shows a different percentile of the distribution, from 1% to 99% (computation performed April 09)

Table 6.9 Historical maxima and minima of one-year Libor rates and ten-year swap rates for several currencies, from 1996/01/01 to 2009/08/17

	1Y Libor Rate		10Y Swap Rate	
	Minimum	Maximum	Minimum	Maximum
CHF	0.303%	3.978%	2.088%	5.029%
GBP	1.160%	8.106%	3.433%	8.449%
JPY	0.064%	1.281%	0.427%	3.664%
USD	0.729%	7.485%	2.296%	7.887%
EUR	1.290%	5.506%	3.131%	6.854%

As a further test to assess the quality of exposure, it is important to mention the comparison between future price distributions with realised valuations, or, in other words, the backtest of predicted exposures. This is one of the regulatory requirements to use EPE and PFE within the Basel II framework for capital calculations. What backtest tries to assess is whether the model used to compute counterparty

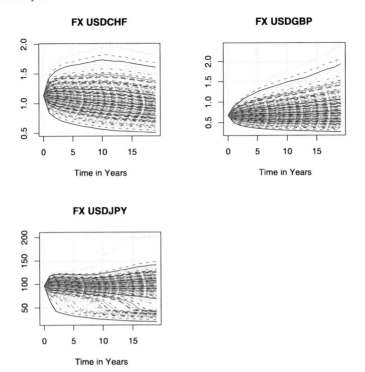

Fig. 6.6 Distributions over 20 years of simulation, of the spot exchange rate for converting USD into the currencies shown. Different lines in the plots correspond to different percentiles (1% to 99%) of the simulated distribution. The plot for JPY indicates a strong skew in the distribution, as is also observed from the data used to calibrate the model (computation performed April 09)

exposure has enough power to predict, in a statistical sense, the future values of a portfolio of transactions. As such, backtesting presents its own challenges.

6.5.2 AMC Error

As valuation algorithm we have chosen AMC. This provides only an approximation of the price of derivatives, and thus of price distributions. Implicitly we have already analysed this point in the previous section when we have compared our price with different benchmarks. In this section we focus on the comparison between price distributions obtained via AMC and closed-form valuation, using the same underlying scenarios.

The AMC algorithm can be implemented in different ways, giving rise, as we also mentioned in Chap. 4, to different sources of errors. We describe here three.

(i) *Choice of observables used in the regression.* As observables determine which are the parameters driving prices, it is crucial to consider carefully which one to use, as an inappropriate choice could lead to unreliable results. The problem of choosing the correct observables is further complicated by the fact that we are interested in computing exposures for large portfolios. As a consequence, the algorithm needs to work in an automatic way, as product types are not known a-priori.

(ii) *Regression error*. Once observables are chosen it is necessary to determine the type of basis functions and their order used for regression.

(iii) *Bundling*. The implementation of the AMC algorithm can be tuned using different bundling algorithms. The size of the bundling can influence results.

To analyse these sources of errors we choose as an example transaction, a 20-year vanilla swap. The example swap has a notional of 100 million USD, and pays semi-annually the 6-month USD Libor, fixed in advance, starting at par.

The price distribution for the swap is computed two ways, one by using the AMC algorithm, and the other by applying the analytical closed-form valuation on each simulated Libor-rate scenario at each time step in the simulation. The AMC valuation results in a time-zero value of 1.18 million USD, as opposed to an analytical value of 0.89 million USD. Figure 6.7 compares the extreme quantiles and median of the two price distributions obtained this way. Based on a notional of 100 million USD and a maturity of 20 years, the time-zero AMC value represents a difference of less than one basis point running from the analytical value.

6.5.3 Numerical Errors

In addition to the approximations embedded in the AMC algorithm, it is necessary to take into account also the typical numerical errors occurring when Monte Carlo

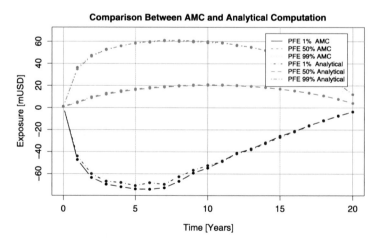

Fig. 6.7 Quantiles of a Vanilla Swap price distribution, obtained through AMC valuation as well as through analytical pricing on each simulated Libor-rate scenario. The swap pays the 6-month Libor rate fixed in advance, in return for a fixed annual rate of 4%, paid semi-annually. Notional is 100 mUSD (computation performed August 09)

simulations are used. The most obvious is related to the number of paths. There are two competing requirements when building a counterparty credit exposure system, accuracy and speed. To obtain reliable results one can be tempted to increase the number of paths significantly. On the other hand to compute a large portfolio this could become quickly unfeasible, even if a distributed grid architecture is implemented (see Chap. 7 for a discussion about the architecture of choice).

Experimentally we have noticed (see also [69]) that about 10,000 simulations provide satisfactory results in a reasonable amount of time. In many cases doubling the number of paths does not improve significantly the quality of the results. Only for payoffs which yield non-zero values in rare cases, such as for example CDSs on small spreads or very out-of-the-money options, a higher number of simulations could be needed.

6.5.4 Approximations: Arbitrage Conditions

When describing our modelling framework in Chap. 2, we pointed out the possibility of introducing simplifications, by allowing the simulation of asset classes independently from the interest-rate process. Doing this enables new models to be added in a modular way and allows calibration for each asset class to be carried out separately. This approach suffers, however, from the existence of arbitrage; for example, in the case of FX, the parity relation at a given time $t > 0$ between foreign currency bonds and the exchange rate is not respected.

To assess the impact of independence, we consider the following example transaction. Fix a time horizon T, say, and consider a contingent claim

$$C_T = \chi_T, \tag{6.59}$$

that simply pays one unit of the foreign currency at T. At each t, C_T should have value $\tilde{D}_{t,T}\chi_t$, the time-t price of a foreign T-bond expressed in the reference currency.

Now consider the option to enter, at time $s < T$, a portfolio that is long C_T and short $\tilde{D}_{s,T}\chi_s$. Because $\tilde{D}_{s,T}\chi_s$ is the no-arbitrage time-s value of the claim, the option should be worthless. If we assume independence, this is not guaranteed, because the FX rate simulation happens independently of the individual bonds (and also because of Monte Carlo and AMC regression error).

As an illustrating example, we fixed s to be the end of the year 2015,[2] and priced the option described above for values T shown in Table 6.10, and for four different currencies. We first computed the prices of the option using the independence assumption—results of this are shown in the third column in Table 6.10. For comparison, we re-did the computation by performing the simulation differently ensuring the interest-rate parity (which entails lifting the independence assumption) that

---

[2]Time of computation is April 09.

Table 6.10 Value, as a percentage of notional, of an option to enter into a portfolio that is long an FX forward and short a bond in the foreign currency. The theoretical value of the option is zero. The table compares AMC values of the option under the assumption of independence between FX rates and interest rates, to the values obtained when that assumption is lifted. The time T is the maturity of the bond and forward. The option's exercise time is fixed at 2015. In the last column we also report the estimated value of the underlying at time of expiry of the option, computed under the assumption of independence between rates and FX. The underlying has theoretical value zero by construction. The notional amount is 100 million units of the payment currency (computation performed April 09)

	T	FX and Rates Independent	FX and Rates not Independent	Value of Option Underlying
USD	2020	<0.1%	<0.1%	−0.27%
	2025	<0.1%	<0.1%	−0.5%
	2030	<0.1%	<0.1%	−0.64%
GBP	2020	3.04%	0.44%	−0.36%
	2025	4.03%	0.56%	−0.6%
	2030	4.15%	1.01%	−0.3%
EUR	2020	3.13%	0.32%	−0.39%
	2025	4.28%	0.47%	−0.55%
	2030	4.26%	0.70%	−0.53%
JPY	2020	4.63%	0.45%	0.0%
	2025	6.46%	0.95%	−0.01
	2030	6.70%	2.78%	0.0%

should make the option value vanish. The results of this computation are shown in the fourth column of Table 6.10. The values, which are not identically zero, represent the AMC error inherent in estimating the value of the option by regression.

The last column in Table 6.10 shows the estimated value not of the option, but of the option underlying, namely the short (resp. long) position in the foreign bond (resp. the claim C_T). As pointed out, this portfolio has theoretical value zero, which is correctly estimated by the algorithm that assumes independence. What this says is that while the option on the error does have value, the error itself vanishes on average, and this is due to the drift correction derived in (2.55).

As expected, the AMC error is smaller than the option value resulting from our modelling framework with the independent assumption. Note that for USD, which we used as reference currency, there is no additional error arising from FX simulation.

The advantage of using the independence assumption is the fact that the FX process can be simulated with a volatility derived directly from observed FX volatility. In order to compute the prices in Table 6.10 without the independence assumption, the process needs to be simulated with a volatility consistent with *both* observed FX volatility and with volatility structures for the individual currencies.

To further assess our simulation for the FX rate, we show as a second example an at-the-money vanilla swap, of maturity 10 years and paying the 12-month Libor rate annually. The swap is in EUR currency, which differs from the reference USD currency and therefore invokes the simulation of the EURUSD FX rate. Results are reported in Fig. 6.8, where the two lines plotted show the modified EPE profiles with and without the independence assumption. It is clear that independence has negligible impact on the computed profile in this case.

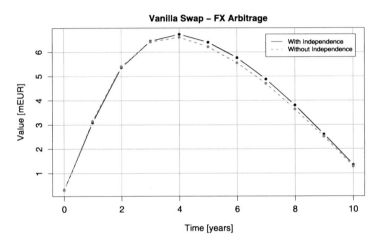

Fig. 6.8 Modified EPE profile for a EUR swap of 10-year maturity and 100 mEUR notional paying yearly. The quoting currency is EUR but the reference currency is USD, so that the FX simulation is invoked during the calculation. The solid line is computed under the independence assumption, while the dotted line results from lifting the assumption (computation performed April 09)

Chapter 7
Architecture

We have described how the AMC algorithm translates into a computational framework that allows systematic counterparty exposure computation of different products. The main result achieved so far is that, within this framework, products are described via their generic features and not their specific definition. This has provided the capability of using functions of financial quantities, which we have called statistics, to define products. As an additional step we have introduced a Portfolio Aggregation Language, PAL, to book trades in the system and, thus, use the analytics in a flexible way. As a result the concept of a new type of product, defined in terms of price distribution of other products, has been introduced. We have called these products super-products, with the most relevant example in this context being the Contingent Credit Default Swap (C-CDS).

All the elements are now ready to build an architecture that has the flexibility of computing the exposure of different products booked in different systems across different asset classes, allowing trading and hedging of counterparty risk.

Designing a suitable architecture is the goal of this chapter. First we consider some very general requirements that should be satisfied by any architecture built to compute and hedge counterparty exposure. Then we describe how to implement different sub-systems with localised functionalities. What we suggest will address typical integration problems large financial institutions generally face.

We will give here only a brief overview of how to design a credit exposure architecture, sketching possible solutions. It is well outside the scope of this book to enter into software engineering details, or give a complete architecture design. On the other hand we believe that implementation issues and architecture design are at the core of any credit exposure system and thus deserve to be at least mentioned and analysed.[1]

---

[1]More information can be found for example in the books by Gamma et al. [47], Evans [43], Gorton [51], and Grune [53].

7.1 Requirements

Large financial institutions are often characterised by having several desks engaged in different businesses, which can potentially use different systems to book transactions. In some cases trading and risk management of positions in new business lines are performed on an individual trade basis, by building new ad-hoc tools. As computing credit exposure is across businesses, this fragmented approach can lead to the impossibility of treating consistently counterparty exposure.

It is necessary to build from the beginning an architecture that copes with different requirements. In general, counterparty exposure systems can be built with two different goals. On one side, they can be risk control systems, used mainly to have a static view of counterparty risk. In this case the users are control functions of the company, with constraints and requirements given to a large extent by regulators. On the other side, a counterparty exposure system can be built with as primary objective to actively manage the firm's counterparty risk. In this case it is the business which is the main user.

Traditionally, risk systems have been built with being conservative as the main requirement. Pricing systems, on the other hand, are designed to be accurate, and the most needed functionality is the ability of computing sensitivities, in order to provide suitable hedges.

Once it has been recognised that a financial company needs to have the capability of pricing and hedging counterparty risk, it is natural to see if the system can serve both communities, Business and Risk. As business requirements are more stringent from the analytical point of view it is also often recognised that it is business responsibility to provide correct measures to risk functions.

The modelling and computational framework we have suggested goes in the direction of pricing. In the following, therefore, we consider an architecture which should first of all be a trading platform, and can then be used also for risk controlling purposes.

7.1.1 Functional, Non-Functional Requirements, and Design Principles

In general, functional requirements, often called also business requirements, are given by users. In contrast, non-functional requirements can be considered as operational specific. Each financial firm building counterparty risk systems will have its own specific requirements. What we consider in the following is only an overview, which helps to highlight how our modelling and design approach provides a sound implementation base.

From a high level perspective an architecture should provide,

(i) *Flexibility*. The architecture should support the exposure computation of a new type of product without additional programming effort.

(ii) *Extensibility.* It has to be able to integrate products from different booking systems in order to have an aggregated view of exposure.

(iii) *Modularity.* The different parts of the system need to be de-coupled. This is achieved by designing well defined interfaces. For example it has to be possible to build new models within the system, without having to change the rest of the code.

(iv) *Scalability.* The system needs to be designed in a way that allows efficient addition of new hardware to compute more transactions.

(v) *Consistency.* All products should use the same set of keywords or statistics, which are all based on the same state of the world.

All these requirements are supported by our computational and modelling framework. For example the way we have defined models gives the possibility of introducing a new model by simply specifying the volatility dynamics. Seen from an architectural point of view, this can considerably simplify the design of the system. Similarly the PAL language gives the possibility of de-coupling the analytical system with various input systems. We will see later, when considering the logical view of the system how other principles can be fulfilled.

7.2 Conceptual View: Methodology

In general, to describe a system architecture several views can be used. We consider here a conceptual, a logical, and a physical view. Often different organisations use these terminologies with different meanings. We call conceptual view the most abstract view of the system. It reflects the underlying methodology used by the analytics, and to a certain extent it represents the functional requirements and the business users' view of the application. The logical view shows the main functional components and their relationships within a system independently of the technical implementation details. The physical view is the least abstract and illustrates the specific implementation components and their relationships. This implementation view is normally owned by IT.

We start by considering the conceptual view. The design we suggest is represented in the enclosed diagram. We have divided the system into three parts, booking, computation, and reporting.

(i) *Booking.* We need to compute exposure at portfolio level. It is therefore necessary to collect all trades that potentially are booked on different pricing systems, but belong to the same counterparty. At this level we need a Portfolio Manager which is responsible for collecting information about trades, netting pools, collateral and collateral agreements.

As we have already mentioned, often large institutions have several independent systems where products of different asset classes are booked. We could have, for example, different systems for vanilla interest-rate products, for exotics, for flow credit derivatives, for correlation products, for FX, and for equity. It is necessary to bring to the counterparty system all information needed

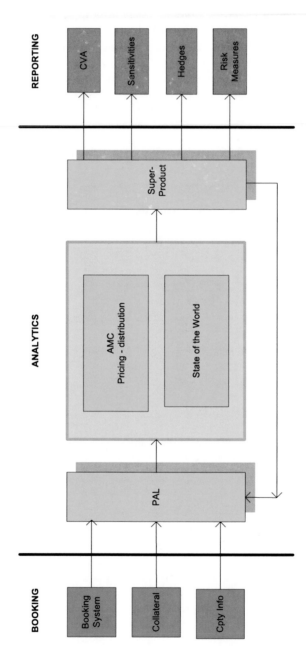

Fig. 7.1 Conceptual view of the architecture

to price a trade. The unification between different systems will be achieved via PAL and PAL translators.

The challenges regarding collateral cannot be underestimated. We will see in Chap. 12 that there are several possibilities to protect from default of the counterparty, and that the process of calling collateral can be complicated.

(ii) *Analytics*. This part of the architecture corresponds to the computational framework we have described in the previous chapters. We have highlighted here the PAL language, used to book trades in the counterparty system; the State of the World, i.e. the scenario generation built taking into account that only Brownian paths can be pre-computed and the appropriate scenarios need to be extracted via statistics; the AMC valuation unit, which provides prices and distribution of prices; the super-products, which feed back the system via PAL.

(iii) *Reporting*. All risk measures such as PFE and EPE, as well as pricing measures such as CVA and sensitivities are computed in the analytics and then reported to various functions.

7.3 Logical View

We consider here a possible logical view, i.e. a view of the main functionalities of the system. In Fig. 7.2 we first show subsystems that could form the basis of the architecture.[2] As main building blocks we have the Portfolio Manager, the Risk Quantification unit, and the State of the World. Each of these subsystems has different components that we describe in the next sections. In addition we need subsystems which interact with users (Risk Quantification GUI), present results (Historical Results Analysis and GUI), calibrate the system with market data (Financial Data Management), provide system diagnostics (System Diagnostics), and run the system on a regular basis (Batch Processor).

7.3.1 Portfolio Manager Components

The Portfolio Manager has to collect information from different booking systems and, using external translators into PAL, present trades to the analytics.

The components of these systems are the following

(i) *Booking System*. This component represent the different systems where trades are booked, generally for valuation purposes.

(ii) *External Translator*. Within the portfolio manager we need a component which translates the trade representation used in each booking system into PAL. We need a different translator for each booking system.

---

[2]We use the standard UML (Unified Modeling Language) [104] to represent sub-systems and components. We will use names starting with "I" to represent interfaces.

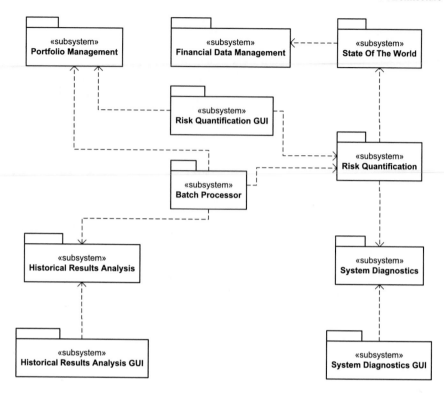

Fig. 7.2 Logical view of the architecture with subsystems

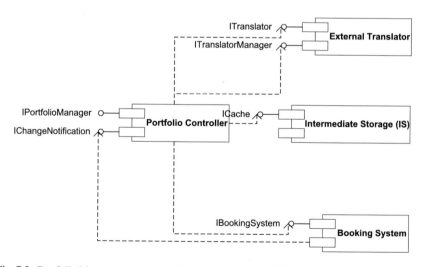

Fig. 7.3 Portfolio Manager components. Names starting with 'I' indicate interfaces

(iii) *Controller*. This is the mediator between components and it interacts between Booking System and External Translator.

(iv) *Intermediate Storage*. As usual, for efficiency reasons, it is necessary to carefully consider the information that needs to be stored.

7.3.2 State of the World Components

This sub-system is responsible for the computation of scenarios. It is therefore characterised by two main components, the Process Simulation, which generates correlated Brownian paths, and the Financial Statistics, which extract the necessary quantities to represent trades. In general a trade-off between computing scenarios at run time, and retrieving them from storage has to be found.

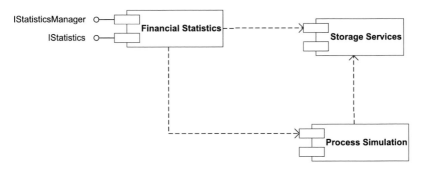

Fig. 7.4 State of the World components. Names starting with 'I' indicate interfaces

(i) *Process Simulation*. This component is responsible for the simulation of the Brownian paths, i.e. the scenarios used to compute price distributions.

(ii) *Financial Statistics*. This component computes the functions needed to build expression trees. It has therefore to communicate both with the trade description and the scenario simulation component, in order to generate the correct statistics.

7.3.3 Quantification Components

The Quantification sub-system is the core of the analytics. We need at this stage the scenarios, computed in an other sub-system, which will be used by the valuation unit to generate price distributions. The quantification controller is the glue of the system. It will invoke the PAL language to drive the computation

(i) *Internal Translator*. This is the PAL booking language we have defined in the previous chapter.

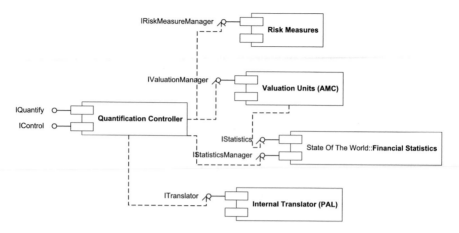

Fig. 7.5 Quantification components. Names starting with 'I' indicate interfaces

(ii) *Valuation Unit.* This is where the pricing is performed on each scenario, as we have mentioned in Chap. 4 the valuation method of choice is AMC. If necessary, however, the valuation unit could include alternative computational methods, such as, for example, analytical functions for vanilla instruments.

(iii) *Risk Measures.* This is the main output of the analytics. We have already described some measures, such as PFE or EPE, used in the industry and we will give more details in Part IV. One of the most important risk measures that need to be produced by the analytics and thus supported by the architecture, is the price of credit risk, or Credit Valuation Adjustment.

7.4 Physical View

It is convenient at this point to briefly consider how a system could be deployed. In general, given the size of the derivative portfolio of financial institutions, a system designed to compute counterparty risk faces a significant computational challenge. The number of trades, simulation paths, and simulation points, usually require to have a grid of machines dedicated to the computation. In Fig. 7.6 we show a possible view. For the communication between nodes we have chosen messaging as it offers loose coupling and reliability. The communication between grid nodes and data nodes depends on the underlying database system, which could be done, for example, via a TCP-IP.[3] Modern database systems have started to use messaging as well.

---

[3]TCP-IP is a network communication protocol.

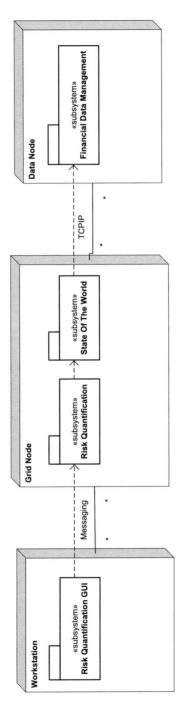

Fig. 7.6 Physical view of the architecture

7.5 Alternative Approaches

With this basic architecture design we can see now how most of the problems we have highlighted in the beginning of the chapter can be solved. Assume for example that different business lines have their own pricing library with their own trade representation. The integration would then consist in translating the specific trade representations into PAL. This is a combination of IT and financial engineering work, as it requires knowledge of the products involved as well as of the booking systems. On the other hand, there is limited required knowledge of the models used either in the pricing systems or in the counterparty system.

In contrast, consider what needs to be performed by credit exposure systems which have a more classical design, where scenarios of underlying risk factors are generated and then pricing functions are called at each scenario and each time step. While this may sound an attractive proposition, as the pricing functions are already implemented in the individual valuation libraries, thus avoiding duplication of work, this approach has a series of drawbacks, which can severely limit the system. Here are some potential issues.

(i) Pricing libraries may differ in age and implementation languages; the process of wrapping these libraries may be more demanding than rewriting the whole library.

(ii) This approach is limited to vanilla trades. If the trade is priced in the valuation library using a Monte Carlo method, it would mean that one would have to implement a Monte Carlo of Monte Carlo method with nested simulations, which is not a feasible solution for a large scale portfolio system.

(iii) Exposure for products which are path dependent cannot be easily computed, as in general scenario history could be in different formats across different systems.

(iv) The risk management system would potentially have to synchronise its releases with the pricing library releases, which could delay usage of new features.

(v) Each new product needs a new implementation or a new integration of a pricing function. This makes difficult to enable the risk system to compute exposure for new products, leaving the risk department behind the business, in the areas where control would be required the most.

(vi) An additional problem is the valuation reconciliation. As we already mentioned a system built in this way is necessarily limited to vanilla products. All other products need to be approximated. The reconciliation of different prices can then become a fundamental issue. While this problem also exists in our proposed design, is solely due to differences in simulation models, and not to any approximation regarding the nature of the trade.

An alternative, which potentially could satisfy the requirements of using existing valuation libraries and also use more sophisticated valuation techniques, is to use AMC valuation whenever needed, and sending scenarios to other libraries, which could then perform close-form valuation. The challenge in this case is technological as relevant amount of information needs to be sent across the network. In addition features such as the concept of super-product we have introduced previously become very difficult to implement, limiting the capabilities of the engine. The design principle of consistency in the valuation approach would also not be satisfied.

Part III
Products

Chapter 8
Interest-Rate Products

In Chap. 4 and 5 we described a generic valuation framework which takes into account the possibility of transactions having early exercise features. In the notation that we introduced there, we represent by $\mathcal{T} = \{\tau_1, \tau_2, \ldots, \tau_n\} \cup \{\infty\}$ the set of times at which the holder of the option may opt to replace the no-exercise portfolio P with time-t value V_t^P, with a different portfolio Q with time-t value V_t^Q ($\{\infty\} = \mathcal{T}^\infty$ indicates no-exercise). The goal of this and of the next chapters is to compute counterparty credit exposure for different types of transactions. As we have already seen in Part II in the first place this means to specify,

  (i) the set of exercise dates $\mathcal{T}$,
 (ii) the type of exercise,
(iii) the cashflows X, paid before exercise, and
 (iv) the cashflows Y, of the exercise portfolio, whose value at time t is V_t^Q.

We will show how to express the product using a payoff description,[1] and we will highlight special features of these products in order to interpret the resulting exposure computations. Throughout we will take $\{T_1, T_2, \ldots, T_n\}$ to represent dates on which some payment is exchanged between one counterparty and another, and $\{t_1, t_2, \ldots, t_n\}$ to represent valuation dates.[2]

We start with interest-rate products.

8.1 Interest-Rate Swaps

An interest-rate swap is a product in which two parties agree to exchange cash-flows related to interest rates in a particular currency. The typical example of such a product is where the Libor rate (fixed in advance or in arrears) is swapped for a

---

[1]To highlight the product features we will not use PAL, but rather functions closer to possible expression trees.

[2]In a continuous time framework this would simply be t.

G. Cesari et al., *Modelling, Pricing, and Hedging Counterparty Credit Exposure,*
Springer Finance, DOI 10.1007/978-3-642-04454-0_8,
© Springer-Verlag Berlin Heidelberg 2009

pre-agreed fixed rate, at a given frequency. It is possible to have swaps where one counterparty pays on a different schedule to the other, and where the notional on which the rates apply vary with time (amortising, accreting, or roller-coaster swaps). In the following we assume that the due cashflows are exchanged on the same day and that the notional is constant. Swaps may be traded in either non-cancellable or cancellable form.[3] The table below summarises the characteristics of these products.

Table 8.1 Product description of different types of swaps. Payments of cancellable swaps are assumed to be in advance. In the capped swap we have not specified if the payment is in advance or in arrears. Floored swaps can be expressed in a similar way. L is the Libor rate, c the annualised fixed coupon, K the cap on the swap, f the penalty paid to cancel the swap, α the time distance between payments, T_i are the coupon dates, and τ_E is the date of exercise

	Swap in Advance	Swap in Arrears	Cancellable Swap	Capped Swap
$\mathcal{T}$	$\mathcal{T}^\infty$	$\mathcal{T}^\infty$	$\{\tau_1,\ldots,\tau_n\}\cup\mathcal{T}^\infty$	$\mathcal{T}^\infty$
Type	no exercise	no exercise	intrinsic exercise	no exercise
X_t	$\begin{cases} 0, & t \neq T_i \\ \alpha(L_{t-\alpha,[t-\alpha,t]} - c) \end{cases}$	$\begin{cases} 0, & t \neq T_i \\ \alpha(L_{t,[t,t+\alpha]} - c) \end{cases}$	$\begin{cases} 0, & t \neq T_i \\ \alpha(L_{t-\alpha,[t-\alpha,t]} - c) \end{cases}$	$\begin{cases} 0, & t \neq T_i \\ \alpha(Min[L,K] - c) \end{cases}$
Y_t	0	0	$\begin{cases} 0, & t \neq \tau_E \\ f, & t = \tau_E \end{cases}$	0

Non-cancellable swaps are never exercised ($\mathcal{T} = \mathcal{T}^\infty$), whereas for the cancellable version, we have specified the dates on which the holder of the cancellability option can opt to exit the swap. The one-off payment to (or due from) the option holder upon cancellation at τ_E is simply Y_{τ_E}, with $Y_t = 0$ for each $t \neq \tau_E$. We have specified an exercise fee of f.

8.1.1 Swaps in Advance and in Arrears

Table 8.2 shows how the payoff of a swap with coupons paid in advance and in arrears would be represented in pseudo-code. Notice also that a capped (floored) swap, in which the floating payments are constrained to be no more (no less) than a certain fixed value, are a simple modification of the payoff we describe (see Sect. 8.1.2 for more details).

---

[3]Cancellable swaps have embedded an option which has economical value. In contrast break clauses give the right to break the contract at market value and, thus, do not affect the price of the transaction. They have, however, an impact for the counterparty exposure valuation (see Chaps. 12 and 14 for more details).

Table 8.2 Payoff description of a vanilla interest rate par-swap paying 6-month Libor rate. The first row represents a swap in advance and 4.44% is the 10-year swap par-rate computed May 6, 2008. The factor 0.5 takes into account that payments are performed on a semi-annual base without considering specific daily count conventions. Note that we receive positive values and pay negative. In the second row we have a swap in arrears. We need to subtract about 15 bps to have the swap at par. This corresponds to the convexity adjustment needed to compute swaps in arrears

	Date	Payoff
In advance:	T_i	0.5 * (INDEX(IR, USD6M, 6M)-0.044)
In arrears:	T_i	0.5 * (INDEX(IR, USD6M, 0M) - 0.0015 -0.044)

The exposure profile of a 10-year USD par-swap paying semi-annually is shown in the Fig. 8.1 below.

Fig. 8.1 USD Vanilla interest rate par-swap paying fix and receiving floating every six months. Notional is 100 USD (computation performed May 08)

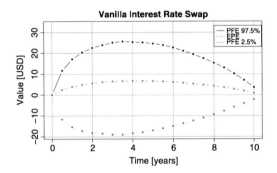

The PFE profile computed at the 97.5% confidence interval has a typical bell shape increasing beginning of the trade and then decreasing as maturity approaches and more cashflows are paid off. It is possible to show that the peak exposure of a vanilla par-swap with constant notional is reached approximately at one third of the profile (see Appendix A for a derivation of this result). The exact position of the peak depends on the value of the yield curve and on its shape. In our example the peak PFE is approximately 25% of notional. In general its value depends on the tenor of the swap (swaps with longer maturity have higher exposure), and on the volatility and shape of the yield curve. Deep in the money or out of the money swaps shows profiles with different characteristics. Note that the profile of a swap in arrears is qualitatively very similar to that of a vanilla swap. The main difference is that the peak is reached in general slightly later in time with higher value.

In general, swaps which receive floating and pay fix show higher exposure than swaps receiving fix and paying floating. This is due to the fact that most yield curves are currently upwards sloping and because there is more volatility on the floating leg than on the fixed leg. This can be seen also in Fig. 8.1 considering the 2.5% quantile of the exposure as this corresponds to the 97.5% quantile of a symmetric trade.

It is also interesting to analyse the relation between EPE and PFE. We can show for example that for a vanilla swap at par, under some simplifying assumptions, the ratio between the EPE and the 97.5% PFE is 20% (see Appendix A).

Amortising or accreting swaps have exposure which can differ substantially depending on the amortising (accreting) notional schedule.

8.1.2 Capped and Floored Swaps

Capped and respectively floored swaps, are swaps in which the floating rate leg is capped or floored to a certain predetermined value. The representation is very similar to the vanilla swap representation as the exercise representation $\mathscr{T}$ is the same and the cashflows X_{T_i} can be described using some appropriate keywords. In Table 8.3 we show the cashflow representation of a capped swap and in Fig. 8.2 the exposure profile of the par swap computed in Fig. 8.1 with floating rate both capped and floored. The value of the swap corresponds to the option premium. Often the fixed rate of the swap is adjusted to take into account the premium and to put the swap back to par.

From the investor perspective capped and floored swaps can be attractive as they reduce or limit payments. From an Investment Bank perspective caps could be used as a device to reduce exposure.

Fig. 8.2 USD interest rate swap capped at 5% and floored at 3%. The value of the optionality is about −4% of notional, assumed to be 100 USD (computation performed May 08)

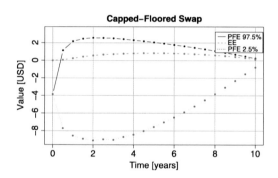

Table 8.3 Payoff description of a capped interest-rate swap paying 6-month Libor rate capped at 5% and floored at 3%. 4.44% is the 10-year swap par-rate

Date	Payoff
T_i	`0.5 * (MAX(MIN(INDEX(IR, USD6M, 6M), 0.05),0.03)-0.0444)`

8.1.3 Cancellable Swaps

A cancellable swap (sometimes called also callable swap) is a swap where one of the counterparties (or both) has the right to cancel the contract at zero value, or at a predetermined fixed amount of cash. The payoff of a cancellable swap is the same as the payoff of a non-cancellable swap. What changes is the set of exercise dates $\mathscr{T}$. We denote long callability the case where the party computing exposure has the right to cancel the swap, and short callability the case where it is its counterparty which has this right. Cancellable swaps can be seen as the combination of a vanilla swap and a Bermudan swaption.

In Fig. 8.3 we consider the exposure of a short and respectively long callability 10 years swap, with callability starting after 5 years.

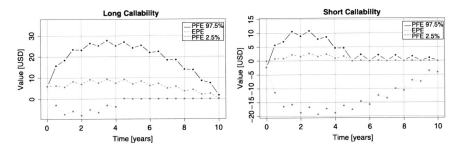

Fig. 8.3 Exposure of a 10 years USD swap, paying yearly fix par rate and receiving semi-annual Libor rate, cancellable annually from year 5. Cancellability (long/short) is seen from the perspective of the counterparty performing exposure computation (computation performed May 08)

If we compare the exposure profile of a vanilla swap with the exposure profile of a cancellable swap, we can notice that long callability has little impact on high quantile exposure. Qualitatively speaking this is due to the fact that a counterparty acting rationally will not exercise its right to cancel the swap, if this has some value. Short callability on the other side will materially impact the PFE, as the counterparty will exercise its right to cancel as soon as the value of the swap, including the option value, becomes negative. This is clearly shown in the right panel of Fig. 8.3. In the first 5 years, where the swap is non callable, exposure follows a profile similar to a vanilla swap. From the fifth year to maturity, the counterparty will call the swap to cancel its obligations. The value of the option, in both short and long callability, is reflected by the value of the swap at inception, which is not zero as it would be in a corresponding vanilla par-swap. As we have already pointed out previously, in practice often the premium is incorporated in the coupon.

8.1.4 Cross-Currency Swaps

Cross-currency swaps involve the exchange of cashflows in two different currencies. We need therefore the simulation of (at least) three risk factors: the two interest rates, and the FX rate. Contrary to interest-rate swaps, there is in general exchange of notional at maturity of the transaction, which exposes the counterparties to movements in the FX rate.

Table 8.4 Payoff description of a vanilla cross currency USD-EUR swap receiving 6 month EUR Libor rate and paying USD fixed rate. 4.44% is the 10 years swap par-rate computed May 6, 2008. The factor 0.5 takes into account that payments are performed on a semi-annual base. The first leg takes into account the cashflows due to the coupons and the second leg the exchange of notional performed at maturity. All cashflows are expressed in the pay (domestic) currency. We have assume notional of one unit of USD corresponding to 0.6 EUR and that there is notional exchange at maturity

Date	Payoff
T_i	0.5*(INDEX(IR, EUR6M, 6M)*INDEX(FX, EURUSD, 0M)*0.6-0.044)
T_n	1-0.6*INDEX(FX, EURUSD, 0M)

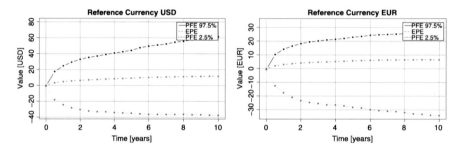

Fig. 8.4 Exposure of a cross-currency 10-years swap, paying USD fix rate and receiving EUR floating on a semi-annual base, with notional exchange at maturity. The two graphs show the exposure in USD and EUR respectively. The notional is 100 USD (computation performed May 08)

When dealing with cross-currency swaps, there is no obvious choice of which currency to use to represent exposure. We can either choose the domestic currency (in our example USD), assuming that we are interested in quoting our potential loss in our own currency, or the receiving currency. If expressed in the receiving currency the loss is limited by the value of the cashflows which has not been paid yet. Note that as for all other products, we can also choose a third currency as reference, as for example an accounting currency or a hedging currency. We will discuss the choice of reference currency in Chap. 12 in the context of aggregation and in Chap. 14 in the context of hedging.

Approximating the max PFE computation is relatively straightforward for cross currency swaps, as most of the exposure comes from the notional exchange at trade maturity (see Appendix A for details).

8.2 Constant-Maturity Swaps and Steepeners

Constant Maturity Swaps (CMS) exchange cashflows which depend on given points of the swap curve. The product characteristics do not change from what we summarised in Table 8.1. What changes is the booking, which has to specify which

swap rate is used. Consider for example a CMS steepener paying the difference between the 10 and 5-year swap rate. It will be described in the payoff language as
follows.

Table 8.5 Payoff description of a steepener, which depends on the difference between the 10 and
5-years swap rate

Date	Payoff
T_i	SWAPRATE(USD, 10Y, 12, 0M) - SWAPRATE(USD, 5Y, 12, 0M) - c

Fig. 8.5 Exposure
comparison between a vanilla
interest-rate swap and a
steepener. The notional is 100
USD (computation performed
August 09)

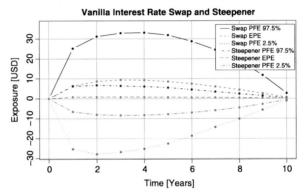

Combinations of different swap rates allow investors to take a view on specific
behaviours of the swap curve. Steepeners are one example where cashflows are computed as difference between two swap rates, e.g. the 10 years and the 5 years. As the
name of instrument suggests, this gives a view on the steepness of the swap curve.
To compute credit exposure (as well as for pricing steepener options) it is important to have an interest rate model with a dynamic which is rich enough to capture
the correlation between different point of the swap curve. In this sense a one factor
interest-rate model with only one driver may not be sufficient. At least a 2-factor
model is needed to get meaningful results. The shape of the steepeners PFE and
EPE profiles is similar to the vanilla swap profiles, with a maximum value reached
approximately in the middle of the life of the trade. It is worth noting, however, that
steepeners have maximum value of the PFE generally lower than vanilla swaps (see
Fig. 8.5). This is due to the fact that the two point of the yield curve and the two legs
are strongly correlated.

8.3 Range Accruals

A range accrual contract is a swap involving the exchange of an exotic coupon,
which is a function of the average presence of a given observable within a specified

range, against a fixed rate leg. In other words, it pays the number of days spent within the range by the observable, divided by the number of observation days.

Let's consider for example a range accrual on USD Libor. Assume it pays every 3 months a coupon in USD of 5% times the average presence of Libor between 4 and 5 %. In a more complex setting, the coupon could also accrue over time with a snowball effect. In this case each coupon could be, for example, the previous coupon times the Libor rate. We will show practical examples of these product in Chap. 11, where we consider Callable Daily Accrual Notes (CDRAN).

Table 8.6 Payoff description of a range accrual paying every three months 5% times the average presence of Libor between 4 and 5 %

Date	Payoff
T_i	0.25 * 0.05* INRANGEFRACTION(IR, USD3M, 3M, 0.04, 0.05)

8.4 Interest-Rate Options

We consider in this section interest-rate options commonly used in the banking in-dustry, that is caplets, floorlets, digitals options, and swaptions. Caplets and floorlets are options on interest rates. In general they are used to construct caps and floors, which are collections of single options. Together with swaps they are used to build capped (floored swaps) such as those presented in the previous sections. Asian op-tions are options which average the index over a certain period. Digitals pay either a fixed amount or nothing depending on whether the underlying is above or below a predetermined level at maturity. Swaptions are options to enter in a swap. They can be physically or cash settled. While for pricing this does not have an impact, in the case of credit exposure computation, a physically settled swaption will show exposure also after expiry of the option when the counterparties enter into the swap.

Most of these products can be priced analytically and therefore classical Monte Carlo approach with analytical pricing at each scenario and each time step could be used. As in the previous sections, however, we describe these products in term of payoff and of our generic framework. The characteristics of these products are summarised in Table 8.7. Using this table and the appropriate keywords, we can express the different products.

We show in Fig. 8.6 the exposure profile of a cash settled and of a physical settled swaption, where the party performing the exposure computation has the right to enter the swap (long callability). In both cash and physical settlement the exposure is the same till expiry of the option. It changes substantially after this date (year 5 in our example). The physical settled swap shows exposure for another 10 years, till the final maturity of the swap, while the cash settled swaption simply terminates after option expiry.

Table 8.7 Interest rate options description. T_s is the option expiry date, L is the Libor rate (caplet) or an average (Asian), A_s is the value at time T_s of an annuity, S is the swap rate, and K is the strike. The formula has to be adjusted for floorlet and for digitals

	Caplet, Floorlet	Cash-settled Swaption	Physical-settled Swaption
$\mathscr{T}$	$\mathscr{T}^\infty$	$\mathscr{T}^\infty$	$\{\tau_E = T_s\} \cup \mathscr{T}^\infty$
Type	no exercise	cash settlement	physical settlement
X_t	$\begin{cases} 0 & t \neq T_s \\ (L-K)^+ & t = T_s \end{cases}$	$\begin{cases} 0 & t \neq T_s \\ A_t(S_t - K)^+ & t = T_s \end{cases}$	0
Y_t	0	0	$\begin{cases} 0 & t \neq T_i \\ \alpha(L-K) & t = T_i \end{cases}$

It is interesting to analyse the 2.5% quantile, as this corresponds to the 97.5% quantile of the reverse trade, which corresponds to swaption, where the counterparty has the right to enter the swap (short swaption). In this case, before option expiry we are not exposed, as the option premium is paid upfront. It may, however become positive if the swaption is physically settled and the two parties enter the swap.

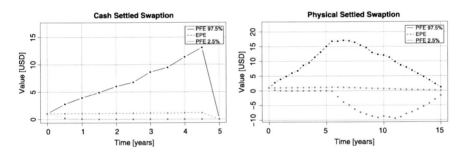

Fig. 8.6 Exposure of a 5 years into 10 years cash settled and physical settled USD swaption. We are long the option, the strike is 3.5% and the holder of the option receives fixed rate. The notional is 100 USD (computation performed May 08)

Chapter 9
Equity, Commodity, Inflation and FX Products

In the previous chapter we considered some of the most common products traded in the interest-rate market. We now turn our attention to standard equity, commodity, foreign exchange and inflation products.

Even if there could be alternatives, we consider inflation and commodity together with equity and foreign exchange products, as the models for these asset classes are in essence the same. For example, a relative return swap on inflation can be computed considering the inflation index,[1] as a particular type of currency. The difference in behaviour comes mainly from market parameters and volatility representation. Compared to equity or FX, inflation shows in general a small volatility of only a few percentage points. On the other side it is the drift, i.e. the difference between the real and the nominal interest rate that dominates the counterparty risk in these products.

9.1 Forwards and Options

We start with forward contracts and with options in their European, Asian, and Bermudan (or American) flavours, together with some slightly more exotic products such as barrier options. These products are important not only because they are some of the most common financial products in the market, but also because they are often part of more complex structures. In the commodity market, for example, it is usual to trade strips of forwards on some underlying with different strikes. Similarly options can be combined in different ways to get specific payoffs, to interpret a specific view on the market. In this section we will focus mainly on examples in which the underlying risk driver is the USDEUR exchange rate, in order to be able to compare the product characteristic rather than the characteristics of the underlying. Our conclusions on these FX examples, however, can be used also for the other asset classes.

---

[1] An inflation index used in the UK is, for example, RPI Retail Prices Index.

G. Cesari et al., *Modelling, Pricing, and Hedging Counterparty Credit Exposure,*
Springer Finance, DOI 10.1007/978-3-642-04454-0_9,
© Springer-Verlag Berlin Heidelberg 2009

In Tables 9.1 and 9.2 we show how these products are represented in our framework. Only Bermudan options have more than one exercise date. In principle exposure for all other products could be then considered in a more classical framework, which does not consider early exercise. Most of these products in their simplest formulation have in addition analytical pricing and, as we will see, their exposure can be approximated in simple ways. This is the reason why counterparty systems targeting these types of products are relatively simple to build, but cannot then be extended easily to cope with more exotic products.

Table 9.1 Forwards, vanilla and Bermudan options description. S_t is the value of the underlying at time t, K the strike price, T the maturity of the product, and τ_i a callable date

	Forwards	Call Option	Digital Put Option	Bermudan Call Option
$\mathscr{T}$	$\mathscr{T}^\infty$	$\mathscr{T}^\infty$	$\mathscr{T}^\infty$	$\{\tau_1, \ldots, \tau_n\} \cup \mathscr{T}^\infty$
Type	no exercise	no exercise	no exercise	intrinsic exercise
X_t	$\begin{cases} 0, & t \neq T \\ S_t - K \end{cases}$	$\begin{cases} 0, & t \neq T \\ (S_t - K)^+ \end{cases}$	$\begin{cases} 0, & t \neq T \\ \mathbb{1}_{S_t \leq K} \end{cases}$	0
Y_t	0	0	0	$\begin{cases} 0, & t \neq \tau_i \\ S_t - K \end{cases}$

Table 9.2 Barrier and Asian options description. S_u denotes the value of the underlying at time u, K denotes the strike price and H denotes the barrier level

	Up & Out Call Option	Down & Out Put Option	Down & In Call Option	Asian Call Option
$\mathscr{T}$	$\mathscr{T}^\infty$	$\mathscr{T}^\infty$	$\mathscr{T}^\infty$	$\mathscr{T}^\infty$
Type	no exercise	no exercise	no exercise	no exercise
X_t	$\begin{cases} 0, & t \neq T \\ (S_t - Kt)^+ \\ \times \mathbb{1}_{\max_{u \in [0,t]} S_u \leq H} \end{cases}$	$\begin{cases} 0, & t \neq T \\ (K - S_t)^+ \\ \times \mathbb{1}_{\min_{u \in [0,t]} S_u \geq H} \end{cases}$	$\begin{cases} 0, & t \neq T \\ (S_t - K)^+ \\ \times \mathbb{1}_{\min_{u \in [0,t]} S_u \leq H} \end{cases}$	$\begin{cases} 0, & t \neq T \\ (\frac{1}{t}\int_0^t S_u du - K)^+ \end{cases}$
Y_t	0	0	0	0

9.1.1 Forwards Contracts

A forward contract is an agreement to buy (or sell) at an agreed date in the future a pre-specified amount of underlying (e.g. shares, currency, oil barrels...) for a pre-specified price.

In Table 9.3 we show how to represent this product in our payoff language. We have chosen the strike to be the current forward price, so as to value the trade at par. The exposure profiles, 97.5% and 2.5% PFE and EPE are shown in Fig. 9.1.

Table 9.3 Payoff description for a forward contract on the USDEUR FX rate struck at 0.65 EUR per USD (par rate) and maturing at T on a unit USD notional. Note that we follow the convention that USDEUR means "USD value in EUR", or, in other words, a USD amount is multiplied by USDEUR to get a EUR amount. In this example we pay EUR and receive USD. The results are expressed in EUR

Date	Payoff
T	INDEX(FX, USDEUR, 0M)-0.65

Fig. 9.1 Exposure of the typical par FX-forward described in Table 9.3. The USDEUR volatility used in the computation is about 10%. On a notional of 100 USD this corresponds to approximately 20 USD PFE on a 97.5% confidence level (computation performed August 08)

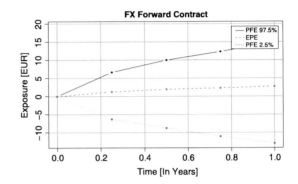

The typical credit exposure profile of a forward shows the maximum PFE value at maturity, where cashflows are exchanged. The exact profile will depend on the drift, of the volatility structure, and on the model used for the volatility. A rough estimate of the max PFE exposure can be obtained by assuming that the FX rate is normally distributed and by multiplying the volatility (assumed to be constant) with the square root of time and the corresponding quantile (see also Appendix A). In this example the 97.5% maximum exposure could be roughly approximated as $FX_0 1.96\sigma\sqrt{T}$. The full PFE profile can be computed in analytical form, if one assume the dynamics of the underlying to be a Geometric Brownian Motion with constant volatility and drift.

When computing exposure it is necessary to choose a reference currency to represent the results. In general several possibilities are available. In some cases there is a natural choice, e.g. we represent exposure in our domestic currency, or in the same currency in which we hold collateral or hedge the counterparty risk, and in some cases the choice may not be straightforward. When dealing with FX products, there are at least two natural candidates, the two currencies of the contract. Depending on the choice the results can be significantly different. For example if one chooses the receive currency to express the results, the maximum PFE will be limited by the notional received at maturity. This is, in fact, the maximum amount we could lose in case of default of the counterparty. On the other side, by choosing the domestic currency (pay currency) as reference, there is no theoretical limit in the exposure. Choosing a third currency as reference, may sometimes complicate the interpretation of the results.

9.1.2 Vanilla and Digital Options

A vanilla European call (put) option gives the holder the right, but not the obligation, to buy (respectively sell) at an agreed future date (exercise date) a pre-specified
amount of underlying for a pre-specified price (strike price). It is interesting to note
that since the price of a long (short) option is always positive (respectively negative), long options will show positive EPE and PFE, while short options will show
zero EPE and negative PFE. For high confidence interval quantiles, the PFE profile
of options and forwards are very similar. As for the forward contracts, in general
options show the maximum exposure at maturity and can be approximated in the
same way we showed in the previous section.[2]

A digital call (resp. put) option will pay at maturity either its full nominal amount
or nothing depending on whether at maturity the underlying value is above (resp.
below) a pre-specified level, or strike. To model these products we need to use the
indicator function, which in our payoff language is translated in INDICATORBE-
LOW, INDICATORABOVE. This will indicate if the underlying is above or below
the given strike.

At maturity there will be scenarios both above and below the strike level. Depending on the level of the confidence level and of the value of the strike, the option
will then return all or nothing. In that sense the PFE can change substantially by
just choosing two different confidence intervals. The behaviour of EPE, which is a
mean, is more stable. For these products it can also be convenient to use other risk
measures such as Expected Shortfall .

In the table below we show the payoff description of a call and put vanilla and
digital option on the same FX USDEUR underlying.

Table 9.4 Payoff description for vanilla and digital European options on the USDEUR FX rate
struck at 0.6 EUR per USD and expiring at T on a unit USD notional

	Date	Payoff
Vanilla Call	T	MAX(INDEX(FX, USDEUR, 0M)-0.6, 0.0)
Vanilla Put	T	MAX(0.6-INDEX(FX, USDEUR, 0M), 0.0)
Digital Call	T	INDICATORABOVE(INDEX(FX, USDEUR, 0M),0.6)
Digital Put	T	INDICATORBELOW(INDEX(FX, USDEUR, 0M),0.6)

9.1.3 Bermudan and American Options

Options with more than one possible date of exercise are known as Bermudan options. Their extension to continuous exercise (the holder has the right to exercise
continuously until maturity) are called American options.

---

[2]For example very out-of-the money options behave differently.

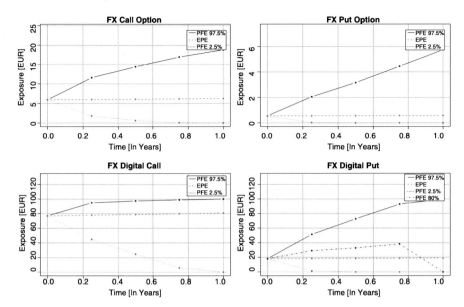

Fig. 9.2 Different FX USDEUR long vanilla and digital options computed on a notional of 100 USD. The corresponding short versions of these product would have negative PFE, if the premium is paid upfront. The results are shown in EUR. The first call option is in the money, as the par-forward is at 0.65. Thus the put option computed with the same strike is out of the money. This explains while the premium for the put and call are different (neglecting the volatility smile), and why the call option has higher exposure than the put (computation performed August 08)

Consider a Bermudan option. For the vanillas of previous sections (see Table 9.1), the payoff coincides with the cashflow X paid at maturity. In the case of a Bermudan option, the holder of the option has the right, at some predetermined dates, to transform the product into a given cashflow Y. The AMC algorithm will decide if it is more convenient to exercise, or to continue and wait for a better opportunity. The payoff description is therefore

Table 9.5 Payoff description of a Bermudan option on the USDEUR FX rate struck at 0.65 and expiring at T on a unit USD notional. T_i are the exercise dates

Date	Payoff of X
T	0

Date	Payoff of Y
T_i	INDEX(FX, USDEUR, 0M)-0.65

The figure below shows the exposure of a structure with two legs, a long European Option and a short Bermudan with the same characteristics. The exposure is practically zero till the Bermudan option is exercised, in which case only the long European call is left generating exposure.

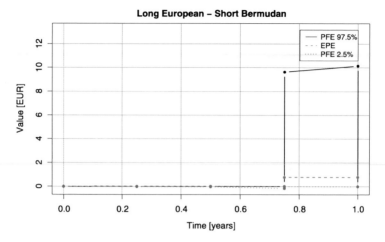

Fig. 9.3 Structure with long European call and short Bermudan call option on FX USDEUR. The option can be exercised every 3 months. All other product characteristics are the same of the vanilla call represented in Fig. 9.2. Notional is 100 USD (computation performed August 08)

9.1.4 Asian Options

Asian options have the same main characteristics of the vanillas we have considered previously. Again the payoff will be described as the value of cashflow X paid at maturity. What distinguishes these products is the fact that the payoff depends on the average of an underlying (e.g. equity or FX) for a specified period of time, instead of a spot value. The payoff description on a FX Asian option is shown in Table 9.6.

Table 9.6 Payoff description for an Asian Call option on the USDEUR FX rate struck at 0.6 EUR per USD and expiring at T on a notional of 100 EUR

Date	Payoff
T	MAX(AVERAGE(FX, USDEUR, 0M, 1Y) - 0.6, 0.0)

In general Asian options show PFE and EPE exposure smaller than the corresponding vanillas. We can see this by noting that the average volatility can be approximated as the spot volatility divided by $\sqrt{3}$ (see for example [62] for more details). The figure below is an example of an Asian FX option on USDEUR.

9.1.5 Barrier Options

Barrier options are extensions of vanilla options which incorporate information on either the maximum or minimum value of the underlying during a specified period of time. We distinguish between:

Fig. 9.4 Asian call option on FX USDEUR. All other product characteristics are the same of the vanilla call represented in Fig. 9.2. We can see that both premium and PFE exposure are smaller that the vanilla example (computation performed August 08)

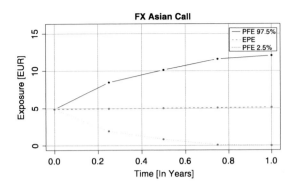

(i) *Up-And-Out (resp. Up-And-In) Call (resp. Put) Options* will pay at maturity the underlying price minus the strike price (resp. the strike price minus the underlying price) if this is positive and if the maximum value of the underlying price during the life of the option has not exceeded (resp. *has* exceeded) a pre-specified barrier level.

(ii) *Down-And-Out (resp. Down-And-In) Call (resp. Put) Options* will pay at maturity the underlying price minus the strike price (resp. the strike price minus the underlying price) if this is positive and if the minimum value of the underlying price during the life of the option has not gone below (resp. *has* gone below) a pre-specified barrier level.

We show in Table 9.7 and Fig. 9.5 some typical examples of FX barriers.

Table 9.7 Payoff description for barrier options on the USDEUR FX rate struck at 0.6 EUR per USD and expiring at T on a unit EUR notional. Respective barrier levels are 70, 58, 58 and 70 EUR cents per USD

	Date	Payoff
Up-And-Out Call	T	`MAX(INDEX(FX, USDEUR, 0M)-0.6, 0.0) * INDICATORBELOW(EXTREMUM(FX, USDEUR, 0M, 1Y, TRUE), 0.7)`
Down-And-Out Put	T	`MAX(0.6 - INDEX(FX, USDEUR, 0M), 0.0) * INDICATORABOVE(EXTREMUM(FX, USDEUR, 0M, 1Y, FALSE), 0.58)`
Down-And-In Call	T	`MAX(INDEX(FX, USDEUR, 0M)-0.6, 0.0) * INDICATORBELOW(EXTREMUM(FX, USDEUR, 0M, 1Y, FALSE), 0.58)`
Up-And-In Put	T	`MAX(0.6 - INDEX(FX, USDEUR, 0M), 0.0) * INDICATORABOVE(EXTREMUM(FX, USDEUR, 0M, 1Y, TRUE), 0.7)`

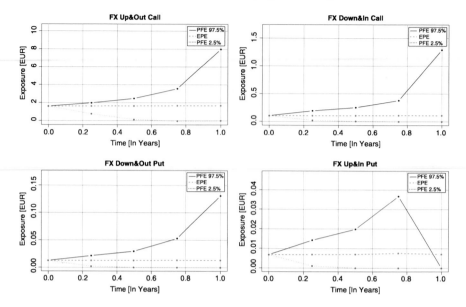

Fig. 9.5 FX USDEUR barrier options examples corresponding to the payoff of Table 9.7. In general these products have smaller premium and exposure that the corresponding vanillas. The exposure profile, is however, very different, as it depends on both the level of strike and barrier. The Up&Out call will pay at maturity the value of the call option (strike 0.6) only if the underlying has not exceeded during the life of the trade the barrier of 0.7. Considering that the forward is at 0.656, this reduces substantially the value of the option. The Down&In will pay only if the FX is below 0.58 during the life of the trade. The put options have similar behaviour. Notional is 100 USD (computation performed August 08)

9.2 Asset Swaps

In this section we analyse the credit exposure behaviour of common asset swaps. These structures generally exchange either the absolute or the relative return of an asset at some fixed predetermined dates. In some cases only the positive (or negative) return is paid. The general characteristics of asset swaps are shown in Table 9.8. These in general are products without callability features.

The example we will analyse will be equity structures as these are the most commonly traded in the market. The results can then be extended to other asset classes.

9.2.1 Absolute Return Swaps

An absolute return swap will pay at a set of pre-defined dates T_i the absolute difference between the current underlying price and its value on the previous fixing date.

Table 9.8 Absolute, Relative Return Swaps, Cliquets and Target Redemption Swaps Description. S_t denotes the price of the underlying at time t and K the strike price

	Absolute Return Swap	Relative Return Swap	Cliquet	Target Redemption Swap
$\mathcal{G}$	$\mathcal{G}^\infty$	$\mathcal{G}^\infty$	$\mathcal{G}^\infty$	$\mathcal{G}^\infty$
Type	no exercise	no exercise	no exercise	no exercise
X_t	$\begin{cases} 0, & t \neq T_i \\ (S_t - S_{T_{i-1}}) \end{cases}$	$\begin{cases} 0, & t \neq T_i \\ \left(\frac{S_t - S_{T_{i-1}}}{S_{T_{i-1}}}\right) \end{cases}$	$\begin{cases} 0, & t \neq T_i \\ \left(\frac{S_t - S_{T_{i-1}}}{S_{T_{i-1}}}\right)^+ \end{cases}$	$\begin{cases} 0, & t \neq T_i \\ (S_t - K) \\ \times \mathbb{1}_{\int_0^t (S_u - K)^+ \mathbb{1}_{u=T_i} du \leq H} \end{cases}$
Y_t	0	0	0	0

9.2.2 Relative Return Swaps

A relative return swap will pay at a set of pre-defined dates T_i the relative performance between the current underlying price and its value on the previous fixing date.

Table 9.9 Payoff description at time T_i of an Absolute Return Swap on the DJ Euro Stoxx with quarterly payments

Date	Payoff
T_i	INDEX(EQ, ESX, 0M) - INDEX(EQ, ESX, 3M)

Table 9.10 Payoff description at time T_i of a Relative Return Swap on the DJ Euro Stoxx with quarterly payments

Date	Payoff
T_i	INDEX(EQ, ESX, 0M) / INDEX(EQ, ESX, 3M) - 1.0

The results for the absolute return swap example described in Tables 9.9 are shown in the figure below, where we assumed that we pay the performance of the stock and we receive the under-performance.[3] From a qualitative point of view, the results for an equivalent relative return swap are similar. At each payment date the performance of the stock is exchanged and the instrument resets. To highlight this behaviour we have sampled the exposure profile one day before and one day after payment date. It is interesting to note that in general the absolute return swap exposure is not centered around zero. This is due to the fact that a stock in general pays

---

[3]In practice this means that the Payoff in Table 9.9 has to be multiplied by -1.

dividends and therefore the discounted expectations of future prices are different from the current stock prices.

Fig. 9.6 Absolute return swap on DJ Euro Stoxx paying quarterly (computation performed August 08)

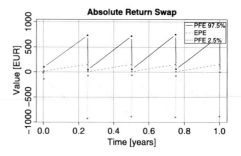

9.2.3 Cliquets

Cliquets pay the positive relative performance of the underlying stock price on a set of dates. It can be seen as a strip of forward-starting at-the-money options, which periodically settles and resets its strike price at the level of the underlying during the time of settlement.

Table 9.11 Payoff description for an Cliquet payment at T_i on the DJ Euro Stoxx with quarterly payments on a unit EUR notional

Date	Payoff
T_i	MAX(INDEX(EQ, ESX, 0M) / INDEX(EQ, ESX, 3M) - 1.0, 0.0)

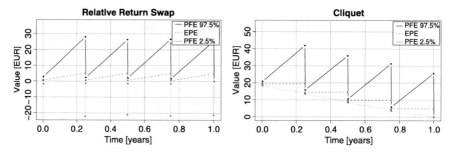

Fig. 9.7 Relative Return Swap and Cliquet on the DJ Euro Stoxx on a notional of 100 (computation performed August 08)

In Fig. 9.7 we compare a relative return swap (as described in Table 9.10) on a notional of 100 where we receive the performance of the stock (and pay Libor), and a cliquet as described in Table 9.11 on the same DJ Euro Stoxx underlying. Cliquets are popular instruments, because they allow to take positions on a stock's future volatility at a lower cost than traditional strategies such as straddles.

9.2.4 Target Redemption Swaps

A target redemption swap can be seen as a strip of forward contracts, with the specific feature that the trade terminates when the sum of the positive parts of the cashflows exceeds a pre-specified barrier level. Target redemption forwards are popular instruments in the equity and FX market because they allow investors to buy stocks or exchange currencies at lower prices than current future prices. We consider here an example on again the DJ Euro Stoxx.

Table 9.12 Payoff description of a 1 year Target Redemption Swap payment on DJ Euro Stoxx, with quarterly payments. Inception date is August 4, 2008 with an index level of about 3200

Date	Payoff
T_i	INDEX(EQ, ESX, 0M) - 3200.) * INDICATORBELOW (ACCRUEDCPFLOATER(EQ, ESX, 0M, 04/08/2008, 4, 3200., FALSE), 2000.)

Fig. 9.8 Target Redemption Swap on DJ Euro Stoxx paying quarterly (computation performed August 08)

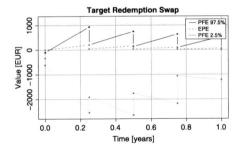

Chapter 10
Credit Derivatives

In this chapter we consider credit derivatives, focussing on loss products. In our framework single name CDSs are just a special case of multi-name CDOs, as the loss dynamics can be described in the same way for both product types. As, however, CDSs have specific features which can be used to introduce more general characteristics of other credit derivatives products, we consider first this type of products. We move then to the classical CDO tranches and show how credit exposure strongly depends on the seniority of the tranche.

10.1 Credit Default Swaps

In its simplest form, a Credit Default Swap (CDS) is a product in which one party, A say, agrees to pay a fee to a counterparty, B say, to get protection on a reference entity (e.g. a bond issued by a third company), should it default. In case of default of the reference entity, A will deliver to B the defaulted bond, and will receive from B the value of the notional of this bond. In other words if R is the recovery rate of the reference bond, the net cashflow from B to A will be Notional $\times (1 - R)$. CDSs come in different other forms and flavours, but this is the typical structure and we will use this to discuss their credit exposure behaviour.

Within our framework, CDSs can be described as interest rate swaps. We can therefore refer to Table 8.1 in Chap. 8 for more details. What differs is the payoff description. In the table below we show how to book a vanilla CDS. The first term describes the loss in a given time interval (tenor), due to the default of the underlying. This is the protection leg. The second term is the protection fee (spread) paid to be protected. This is the payment leg. As in case of default of the underlying no further payment is due, we need to multiply the spread by (1-CreditEvent).

In the figures below we show the credit exposure for different CDSs. The first panel on the left is a CDS where exposure is computed from the perspective of the protection buyer (long protection). The spread of the underlying bond is about 60 bps, corresponding to a default probability of about 1% per annum, assuming a recovery rate of 40% and notional $100. In addition to the usual statistics we have

G. Cesari et al., *Modelling, Pricing, and Hedging Counterparty Credit Exposure,*
Springer Finance, DOI 10.1007/978-3-642-04454-0_10,
© Springer-Verlag Berlin Heidelberg 2009

Table 10.1 Payoff description of a credit default swap, starting at T_0 and paying a fee s to receive default protection on a bond B. We assume that payments are performed every six months. The characteristic of the bond B are described in terms of spread and recovery rate

Date	Payoff
T_i	CREDITLOSS(B, 0M, 6M) - s * 0.5 * (1-CREDITEVENTS(B, 0M, T0))

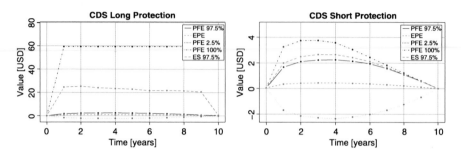

Fig. 10.1 CDS exposure. The reference entity is a bond with spread 60 bps, and recovery rate 40%. Notional is 100 USD (computation performed May 08)

computed the 97.5% expected shortfall and the 100% quantile, that is the maximum loss which can occur. We can see that the PFE at 97.5% confidence interval is about $2.5 million. On the other hand the maximum amount one could potentially lose is much higher, reaching $60 million. The probability of losing this large amount is however very small. This behaviour is typical of products with bi-modal price distributions, where one of the mode is driven by the default of the underlying. Only if the quantile at which exposure is observed is high enough, is the default mode observed. For these products, PFE is probably not the appropriate measure of risk and a better view is given by the mean of the tail, i.e. the expected shortfall.

The second panel on the right side shows a short protection CDS. In this case the maximum amount which can be lost is the value of the payment leg.

10.2 Collateral Debt Obligations

A CDO (Collateralised Debt Obligation) is a product designed to offer loss protection on a portfolio of names within certain predefined limits. Consider two counterparties, A and B, and assume A buys protection on a portfolio of say 100 names equally weighted. As defaults hit the portfolio, B will compensate for losses starting from a given threshold k_A, and not exceeding the level k_D; k_A is called the attachment point and k_D, the detachment point. Depending on the attachment points, CDO tranches are usually classified as,

– Equity tranches, with the attachment point k_A equal (or very close) to zero. The main characteristics of these tranches is that any default hitting the portfolio, generates a loss in the tranche.

- Mezzanine tranches, with attachment and detachment point relatively low. In general they correspond to a rating in the BBB region.
- Senior tranches. These tranches are more senior than equity and mezzanine and correspond in general to a rating of AA.
- Super-Senior tranches. These are the most senior tranches in the structure and corresponds in general to a AAA rating. The probability of having a loss hitting the tranche is extremely low.[1]

A CDO tranche, where we swap protection for a given fee, can be expressed in our payoff language as follows.

Table 10.2 Payoff description of a CDO tranche on a portfolio P with k_A attachment point and k_D detachment point, and paying a protection fee s

Leg	Date	Payoff
Protection Leg	T_i	MAX(CREDITLOSS(P, 0M, T0)-k_A,0) - MAX(CREDITLOSS(P, 6M, T0)-k_A,0) - MAX(CREDITLOSS(P, 0M, T0)-k_D,0) + MAX(CREDITLOSS(P, 6M, t0)-k_D,0)
Fee Leg	T_i	0.5 * s *((k_D-k_A)-MAX(CREDITLOSS(P, 0M, T0)-k_A,0) - MAX(CREDITLOSS(P, 0M, T0)-k_D,0))

This reflects the fact that the tranche loss can be written as a long and a short option on the portfolio with different strikes. With the notation of Chap. 3 this can be written as

$$\Pi_t = (L_t - k_A)^+ - (L_t - k_D)^+,$$

which is reflected in Table 10.2.

In the figures below we show exposure of different CDO tranches on a homogenous portfolio of $1 billion, where all the names are equally weighted and characterised by a spread of 100 bps and a recovery rate of 40%.

Equity or junior tranches show exposure especially at the beginning of the life of the trade. High-spread names in the underlying portfolio, in fact, will default first, possibly reducing the total average spread.

It is interesting to note that senior tranches show substantial counterparty exposure even if defaults of the underlying portfolio do not necessarily hit the tranche. One of the reasons is that as defaults happen, the tranche becomes more junior, while the protection fee remains unchanged. This creates an unbalance in the structure.

In Fig. 10.3 we show the PFE of a 30–60 tranche computed with various levels of correlation using as reference a basket of names with spread of 100 bps. We show in the same picture the PFE of the same tranche on iTraxx. This index, made of

---

[1] In the above rating description we have assumed CDOs to be synthetic. During the 2007–08 credit crisis, it became clear that rating of cash CDOs had different implications.

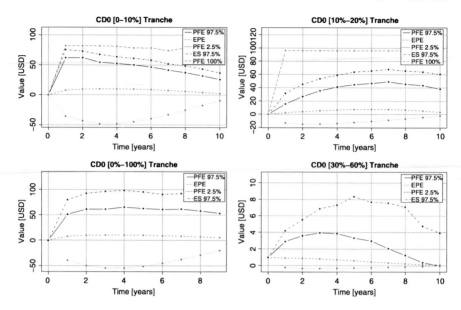

Fig. 10.2 CDO exposure for different tranches paying par coupon. The reference portfolio is constituted by names having the same spread of 100 bps, recovery rate 40% for a total amount of 1000 USD (computation performed May 08)

125 names with spreads ranging from 20 to about 800 bps, has a weighted average spread of about 100 bps (at the time of writing).

Fig. 10.3 CDO PFE exposure of a 30–60 tranche with various level of correlation on an underlying portfolio with spreads of 100 bps. For comparison we also show a 30–60 tranche on iTraxx. The notional of the underlying portfolio is 1000 USD

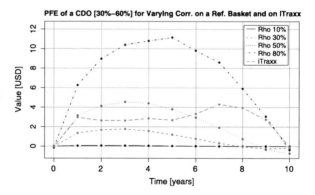

Chapter 11
Structures

We consider now some structured products, which show interesting credit exposure characteristics. As we will see these examples are not necessarily complex to model. They are, however, interesting to examine, as they are commonly used in the industry, and they have features which may only appear in the context of credit exposure. In general, the complexity of these transactions is in the structure itself, and the challenge lies in understanding *what* to model, rather than *how* to model. We will give here only a brief overview of the products and of their structures, without entering into details.

11.1 Sinking Funds

Consider an organisation needing to raise capital. It could be, for example, a municipality wanting to build a school or a hospital, or subsidising mortgages to help some categories of people. One possibility to raise capital is to issue a bullet bond, i.e. to issue a non-callable bond which repays the notional at maturity without intermediate capital re-payments. A bullet bond can be attractive for some investors, but it has the main drawback of bringing substantial counterparty risk. In the case of default of the issuer, the full notional is lost. Seen from another perspective, an inappropriate debt management could bring the issuer to default. Re-payment failures suggest that this financial instrument is not always appropriate to all institutions. For this reason in some countries, institutions such as local authorities can issue bullet bonds only if accompanied by debt amortising swaps, unless a special *sinking fund* is created.

A sinking fund is a method by which an institution sets money aside over time, in order to be able to re-pay the notional at maturity. More specifically, it is a fund into which money can be deposited, so that over time it can be used. From the economical perspective sinking funds have the same behaviour of amortising swaps, but they are more attractive as the amount invested in a sinking fund can also be used for purchasing assets with higher returns.

G. Cesari et al., *Modelling, Pricing, and Hedging Counterparty Credit Exposure,*
Springer Finance, DOI 10.1007/978-3-642-04454-0_11,
© Springer-Verlag Berlin Heidelberg 2009

Consider the following example where party A issues a bullet bond and enters into a back-to-back swap with party B, combined with a sinking fund.

Table 11.1 Typical termsheet of a swap combined with a sinking fund. Party A could be, for example, a municipality, and party B could be a financial institution

Party A	Party B
Pays floating on amortising schedule starting from 500 mEUR and decreasing/amortising 10 mEUR every coupon day (6 months). Pays 10 mEUR in sinking fund semi-annually.	Pays 3.3% semi-annually on 500 mEUR. Pays 500 mEUR at maturity.

The exposure profile seen from party B's point of view is represented in the figure below. The interesting feature is that after a given point in time, exposure is always negative. This reflects the fact that it is party B that has to pay the full notional at maturity, while party A pays the notional on an amortising schedule into a sinking fund. In this way party A is able to repay investors at maturity of the bond. The deposit in the sinking funds can be invested for example in treasury bonds.

Fig. 11.1 Exposure of a swap with sinking fund seen from party B on a notional of 500 EUR (computation performed April 09)

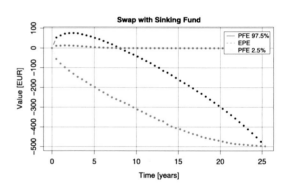

From party B's point of view, credit exposure can be tuned by changing the amortising schedule, and thus modifying the zero cut-off point. In this example the cut-off point is after about seven years. This can be crucial from a credit risk perspective, if there is no appetite for long maturity exposures. Note that the cut-off point will also depend on the volatility of interest rates.

11.2 Accelerated Share Re-Purchase

In a typical Accelerated Share Repurchase (ASR) structure, counterparty A purchases in one block from counterparty B a substantial amount of its own shares, and possibly retire them from the market. At the same time, over a period ranging between three and nine months, counterparty B buys from the market the share of counterparty A. Counterparty A enters into a forward contract with B and agrees to pay to B at the end of the re-purchased period, the difference between the initial price and the volume weighted average price (VWAP) of its own shares.

The ASR program we have described, is a way that can be used by a firm to buy back its own shares from the market without an open announcement. In general, a company enters in an ASR typically when the management of the firm thinks that the share price is traded below what is considered its *fair* value and no better investment opportunities are available.

To better understand the structure and final payoff of an ASR transaction, consider the following example. At the beginning of the trade, party A pays 100 mUSD to party B and party B borrows and delivers 1 mm shares of A's common stock (assuming the share price to be 100 USD). This provides an immediate reduction in A's share count on the market.

To deliver the shares to A, party B shorts 1 mm shares of party A's stock, and then, immediately, begins to unwind the short position through regular purchases from the market typically over several months. The daily purchase amount is entirely at party B's discretion and is chosen in a way to avoid disturbing the market.

At maturity parties B and A will settle the trade typically on a price corresponding to the Volume Weighted Average price of the common stock over the period from the trade date to the final averaging date.

Party B has the option to choose the final averaging date from a set of dates, usually this set is an interval from a predetermined date to the expiration date.

Table 11.2 Typical termsheet of an ASR program

Time	Party A	Party B
Day zero	Pays 100 mUSD	Borrows and delivers shares equivalent to 100 mUSD
Maturity	Difference between initial price and Volume Weighted Average Price	

Counterparty exposure (e.g. from party B point of view) raises from the forward position. It is relatively complicated to compute, as the purchase schedule is discretionary. What is straightforward is to have approximations. A typical way is to compute a forward with Asian characteristics. The exposure profiles for the transaction we have shown above is given in the picture below.

There are benefits that explain why ASR structures are used; (i) they can be structured in very flexible ways, (ii) they are relatively inexpensive compared with other buy-back approaches, and (iii) the share count in the market can be reduced immediately.

Fig. 11.2 Exposure of an
ASR on a notional of 100
USD (computation performed
April 09)

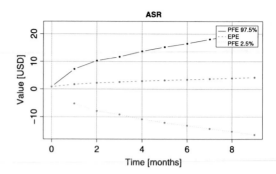

11.3 Callable Daily Accrual Notes

Daily range accrual notes are structured products that pay a coupon whose value accrues with the number of days a given index stays within a certain range. Daily range accrual notes can be structured with or without an embedded call option. In the callable version (CDRAN), the issuer can redeem the note at certain predetermined dates, i.e. it has the right to pay back to investors the bond before maturity.

Both callable and non-callable daily range accrual notes have higher return than conventional bonds. This is due to the embedded optionality these products have. In the non-callable form the investor sells for every day of the life of the product an option with strikes corresponding to the barriers on the index. The issuer will either pay the coupon or nothing, i.e. the payoff is like a digital option. In the callable form, the investor writes an option to the issuer to enable to call it. The issuer pays for this option, thus enhancing the potential return.

CDRAN's maturities are usually between 1 and 10 years. The index used for daily reference is in general the 3 or 6 months Libor rate, but it can be also a spread between two swap rates (e.g. the difference between the 30y and 10y USD rate), or an FX rate or a stock price.

Consider as a practical example the note described in Table 11.3. A typical investor will have the view that the 3 m USD Libor will stay within 1% and 7% for a given "Interest Rate Period". If the index will stay within the range for the whole period, the note will pay a coupon of 6%, which is substantially higher than the typical coupon of a Treasury bond.

If it is not called by the issuer, a CDRAN pays at maturity the notional. During its life, however, the value of a CDRAN can go below par (as it does for all products that depend on stochastic quantities). Potential values of the CDRAN of our example are shown in Fig. 11.3. We can see that if from one side this structured note can potentially achieve higher return, it also brings additional risk. Consider in fact the 2.5% quantile. It shows that during the life of the transaction its value can go substantially below par. In addition, as we can see from the 97.5% quantile, the call feature limits the potential value of the note.

Table 11.3 Typical termsheet of a CDRAN

10 year USD denominated Callable Daily Range Accrual Note with Interest linked to the USD 3 month LIBOR. The Notes are callable by the Issuer after 5 year and quarterly thereafter

Issuer	Party A
Issuers Ratings	AA+ S&Ps / Aa2 Moodys / AAA Fitch
Status	Senior Unsuborbinated
Principal Amount:	[USD 100]
Launch Date	today
Issue Date	today
Maturity Date	10 Years from Issue Date
Interest Amount	(Coupon Rate × (N/365) × Principal Amount −5%)
N	the number of calendar days in the Interest Period that the 3 month USD LIBOR rate is within the following range
Range	[1%, 7%]
Coupon Rate	11%
Early Redemption Option (Call)	The Issuer may redeem the Notes, in whole but not in part, on each Interest Payment Date commencing on one year after issuer date The note holder will be entitled to any Interest payments due on the Early Redemption Date.
Early Redemption Date	If the Notes are called, the Interest Payment Date in respect of which the Early Redemption Option is exercised.
Optional Redemption Amount	100% of the Aggregate Nominal Amount

Fig. 11.3 Exposure of the CDRAN described in Table 11.3. Note that today's CDRAN value is not at par (computation performed April 09)

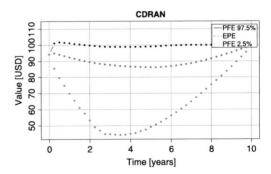

11.4 Call Spread Overlays

Call Spread Overlays (CSO) are typical equity trades that usually involve selling to a counterparty an American call option at a given strike, and buying a European call option at a higher strike from the same counterparty. Often CSOs are structured together with convertible bonds.

Convertible bonds are a type of bond that can be converted into shares of the company issuing the bond, at a predefined ratio. These securities can have several features. We consider here a particular structure, where the issuer of the convertible is interested in increasing the conversion price and, at the same time, retain the maximum upside before having the bond converted into shares. This can be achieved by overlaying to the convertible bond a CSO transaction.

When a CSO is structured together with a convertible, the strike of the American option, which is lower than the European option, is set at the conversion trigger of the bond. If the bond is converted, the American option is also exercised. The upper strike European option usually has maturity between 30 and 180 days longer than the lower strike American option.

Table 11.4 Typical termsheet of a CSO

Counterparty	Party A
Issuer of Convertible Bond	Party A
Buyer	Party B
Underlying	Shares of company A
Underlying value	100
Option 1	Short American call option
Option 1 strike	125
Option 1 maturity	5y
Option 2	Long European call option
Option 2 strike	160
Option 2 maturity	5y6m

Fig. 11.4 Exposure of the CSO described in Table 11.4. We have performed the computation for a notional of 100 USD (computation performed April 09)

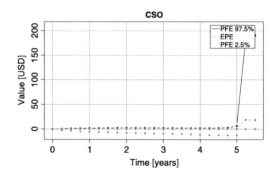

As we can see from Fig. 11.4 the maximum PFE is driven by the long European option at maturity. On the other hand during the life of the CSO trade the EPE profile, and thus the credit charge, are reduced by the netting effect coming from the short American option. If, however, the American option is exercised, the unbalanced European option will attract exposure. To further complicate the assessment of credit exposure is the fact that the underlying shares of a convertible are those of the company issuing the security, i.e. of the counterparty. From the credit exposure point of view this is a right-way risk case, i.e. the default of the counterparty will affect the value of the exposure. We will discuss this topic in more details in Chap. 13.

Part IV
Hedging and Managing Counterparty Risk

Chapter 12
Counterparty Risk Aggregation and Risk Mitigation

In the previous chapters we have considered credit exposure of single transactions. We examine now how to aggregate these exposures at counterparty level and then how to control and manage the risk from a portfolio perspective. This is where the real challenge starts and where it becomes clear why a robust modelling framework is necessary. To obtain a portfolio view it is necessary in fact to calibrate models and to compute products of different nature in a consistent way. In a classical Monte Carlo framework, where exposure is computed in two distinct steps, i.e. first by generating scenarios and then by pricing (using analytical pricers or suitable approximations), this commonality is achieved by using the same consistent scenarios across products.[1] In our framework where scenario generation and pricing are linked together, the scenario consistency is embedded in the underlying pricing model. The hybrid product we need to value taking into account all stochastic drivers in a consistent way, is the given portfolio of transactions.

After having computed exposures it is necessary to find ways of managing the counterparty risk effectively. The most straightforward way of mitigating it is to impose limits on transacted notional amounts, which effectively restricts the mark-to-market value of the transaction. These limits, often based and compared with the value of the PFE, depend in general on the quality of the counterparty. Limits are designed to reflect the credit risk appetite of a company towards its counterparties, as well as to control exposures with a given geographical region or a business type.

This approach translates into a limitation on the amount of business that a firm is prepared to do with a given counterparty. A more active, and still standard practice is to require that the counterparty post collateral, whenever the mark-to-market value of the transaction is seen to increase. Such an agreement allows the profit-making party access to the collateral in the event of counterparty default. If the process is managed properly, the only risk remaining is the close-out risk, which can be

---

[1] Sometimes in the industry and especially in the credit community, the concept of scenario consistency is pushed to unrealistic goals, as it would require data (e.g. correlations) which are not available in the market. In our opinion the real challenge is to find the right balance between simplicity of treatment and consistency.

G. Cesari et al., *Modelling, Pricing, and Hedging Counterparty Credit Exposure*,
Springer Finance, DOI 10.1007/978-3-642-04454-0_12,
© Springer-Verlag Berlin Heidelberg 2009

significantly less material than the PFE over the whole life of the transactions, thus reducing limit utilisation.

One unattractive feature of collateral agreements is the need to post potentially large amounts of collateral upfront and being able to value it in a timely way. Furthermore, when a market move is large enough to necessitate a call for a substantial amount of extra collateral, the client relationship with the counterparty can be put at risk.[2]

In general it is a Credit Risk Control function with dedicated Credit Officers that has the task of choosing the limit structure, decide the amount of collateral to require and monitor PFE (or close-out risk) against it. We refer to this activity as to *Risk Control*, and we consider it intrinsically static. In contrast *Managing Credit Risk*, i.e. pricing and hedging counterparty credit risk with other financial instruments during the life of the trade, is a typical business and front office responsibility, and has a dynamic and trading flavour.

In this chapter we focus on how exposures are aggregated and controlled. Crucial elements of this analysis are the choice of the measure which will serve to assess the riskiness of a position, and, once exposure is aggregated, the possibility of allocating it back to individual components, such as different lines of businesses. We then analyse the most common features in collateralised agreements and consider how to compute exposure and close-out risk for these types of contracts. We do not consider here trading and actively risk managing exposure, as this will be the topic of Chap. 14.[3]

12.1 Risk Measures

Choosing the correct risk measure is crucial. This not only from a theoretical perspective (see for example the seminal paper by Artzner et al. [6] or for a more comprehensive approach the book by McNeil, Frey, and Embrechts [78]), but also from a very practical perspective. The 2008–09 credit crisis clearly shows the importance of finding appropriate measures to monitor the risk of financial institutions. We focus here on some practical aspects.

As we have seen throughout this book, Potential Future Exposure (PFE), i.e. a high level quantile of the price distribution, is the typical measure used to characterise credit exposure in the financial industry.

$$\text{PFE}_{\alpha,t} = q_{\alpha,t} = \inf \{x : \mathbb{P}(V_t \le x) \ge \alpha\}, \tag{12.1}$$

---

[2]During the 2008 defaults it has been become clear that legal and collateral valuation problems can be the major issues when trying to access collateral. In addition large margin calls can accelerate default of a troubled counterparty.

[3]For overviews of counterparty exposure management see for example Arvanatis & Gregory [8], Canabarro & Duffie [22], Canabarro et al. [23], Duffie & Huang [38], Gregory [52], Lomibao & Zhu [75], Pykhtin [87], Pykhtin & Zhu [88], Sorenson & Bollier [100], and Tang & Li [101].

where α is typically 95%, 97.5%, or 99%, depending on the application and usage for which exposure is computed, and $\mathbb{P}$ is a given probability measure. An alternative measure to PFE is the expected shortfall (ES), which has the desirable property of being coherent (see for example [33] and [7]).[4] It is defined as

$$\mathrm{ES}_{\alpha,t} = \mathbb{E}\left(V_t | V_t \geq q_{\alpha,t}\right), \tag{12.2}$$

with the expectation taken in the $\mathbb{P}$ measure.

We could think that the quantile measure answers the following question: *what is the best that can happen if an α% confidence level event occurs?* This definition would miss losses at the more extreme tails. Expected shortfall, on the other hand, corresponds to a different question: *what is expected to happen if an α% confidence level event occurs?* This seems a more natural view point, as it can give a sense of tail risk.

Consider for example a credit derivative. This product presents a bi-modal price distribution. With high probability its value is relatively small, but in certain scenarios, i.e. when the reference entity defaults, which in general happens with very small probability, its value, and thus its exposure can be very high. In case of default in fact, a CDS will pay a fraction of the notional, depending on the recovery rate. In normal circumstances, i.e. in general when the transaction is approved, this event is outside the chosen PFE confidence interval and thus could be completely ignored. On the other side expected shortfall, being the average of the tail distribution, gives a better insight of future risks (see also Chap. 10 for additional examples).

As we have said, ES is a superior risk measure to PFE in the sense that it is coherent. However, the market standard for both counterparty risk and market risk is to use a quantile-based method (PFE or VaR). It is therefore natural to try to find a coherent measure that would give the same risk number as PFE when the portfolio is normally distributed, but at the same time would better account for tail risk when this is not the case. One possibility is to find α such that $ES_{\alpha,t} = q_{\alpha,t}$ for a normal distribution. Indeed, it is easy to value the expected shortfall of a normal distribution, and to show that a 99% (resp. 97.5%) confidence level for PFE or VaR corresponds to a 97.4% (resp. 93.6%) confidence level for expected shortfall.

An other fundamental measure used in the financial industry is the Expected Positive Exposure, EPE. It is defined as the mean of the positive part of price distribution,

$$\mathrm{EPE}_t = \mathbb{E}\left(V_t^+\right). \tag{12.3}$$

This measure is used in practice for hedging counterparty exposure and for computing risk weighted assets and capital. As we will see in Chap. 14 the reason for this choice is that in order to protect against default of a counterparty using credit derivatives, it is necessary to buy a CDS with a notional related to the EPE profile. To achieve this goal, however, the EPE has to be computed under the pricing measure. In Chap. 14 we will also see that (12.3) can be used only if the price distribution V_t

---

[4]We will use coherent measures also in the next chapter, where we discuss stress testing.

is independent from the numeraire and from the default time of the counterparty. If these assumptions are not satisfied a more general formulation of EPE, which we call modified EPE, has to be used

$$\text{EPE}_t^{mod} = \frac{1}{D_{0,t}} \mathbb{E}\left(\frac{V_t^+}{N_t}\middle| \tau = t\right),$$ (12.4)

where N_t is the numeraire, $D_{0,t}$ the cash bond, and τ the default time.

Sometimes it can be interesting to analyse the risk seen from the counterparty point of view. The relevant measure is then the so-called reverse EPE, computed as the mean of the negative part of the price distribution.

$$\text{Reverse EPE}_t = \mathbb{E}\left(V_t^-\right),$$ (12.5)

where $V_t^- = \min(V_t, 0)$. Reverse EPE is used to compute the so called *debt valuation adjustment* (DVA), which corresponds to the CVA seen from the counterparty point of view.[5] Clearly we can also define a modified reverse EPE.

12.2 Choice of Measure

From the pricing theory we know that the measure of choice to be able to build a replicating portfolio is the pricing (called also risk-neutral) measure. This is the measure we have used in the previous chapters of this book and it is consistent with the view that we want to *price* credit exposure.[6] In Chap. 14 we will consider in details what this means in practice. We are here, however, in a different framework. Our goal in this chapter is to measure counterparty risk and not necessarily to price it. In the classical Monte Carlo framework, this means that scenarios should be generated in the *real* measure, while pricing at each scenario is performed in the risk neutral measure. Or more generally, that underlying scenario generation and pricing are two distinct steps, which need to be considered separately even in the way their underlying stochastic processes are measured.

We have seen in the previous chapters, however, that as soon as pricing is not trivial, it can be convenient to use AMC schemes to compute exposure, which use the same scenarios for both projecting future states of the world and pricing.

In certain situations, involving controlling risk and computing capital, this can be problematic. Without entering into details, we consider below some of the issues involved.

(i) Within the Basel II Accord, capital can be computed using the so called Exposure At Default (EAD), i.e. the EPE measured at one year horizon taking

---

[5]Note that in our definition Reverse EPE would have a negative value. If we want to see it as the exposure seen from the counterparty point of view we should multiply its value by -1.

[6]Note that there is an infinity of pricing measures, each related to a different choice of numeraire. Within our framework, we have, however, to choose one, in order to be scenario consistent.

into account some specific rules. To be able to use this measure, exposure has to pass backtesting requirements, discussed and agreed with regulators. These requirements are not necessarily in line with pricing requirements.

(ii) For pricing and hedging purposes models need to be calibrated daily to market prices. This implies that a market view is immediately incorporated into pricing. Even if EAD is computed under a pricing measure, this is not always a desired behaviour to pass backtest. Backtesting, in fact, involves comparing projected exposures with realised valuations over a certain period of time. This needs to calibrate models with parameters which could be different from the actual market view.

(iii) Risk Control functions can be interested in conservative measures that are stable over time, or, in other words, in generating scenarios which span economical cycles rather than replicate the market.

All these problems do not have simple and unique solutions. We will see in the next chapter that one way of tackling the problem of measure of choice when control issues need to be considered, is to first simulate and price under the risk neutral measure, and then apply a change of measure to the price distribution. This approach, however, is not always sufficient to solve all the problems at hand.

12.3 Portfolio Risk Aggregation

In principle, for individual transactions, counterparty credit exposure computation could be performed in isolation, i.e. each front office trading system could compute credit exposure independently using different model libraries. Credit exposure is, however, relevant only at counterparty level, and has to be computed taking into account an aggregated view. This is a challenging point as often, financial institutions do not have a common computational platform across asset classes. Each desk tends to develop its own modelling library in isolation and sends to common risk systems only the present value of transactions and their sensitivities. As we have seen in this book, computing counterparty exposure has to be considered in a different way.

To make these points clear, consider as a very simple example the portfolio of two positions, a call and a put option on two correlated stocks and assume to be interested in the PFE of this portfolio. As we have seen in the previous chapters, a long position in a call option will reach its maximum exposure when the value of the underlying stock increases. It is in these scenarios, in fact, that a call option is in the money. Note that if the position is short, exposure would be zero, as the premium of an option is in general paid up-front. For the put option the situation is different. A long put is in the money when the stock value decreases. Taken in isolation the two contracts will in general show significant exposure. The portfolio of the two options shows, however, a completely different behaviour, as the scenario consistency approach dictates that we cannot assume at the same time high and low levels of stock prices.

12.3.1 Reference Currency

Before starting aggregation it is necessary to choose a reference currency to represent the value of all transactions. This choice is not obvious. Even for single transactions it is not always clear in which currency credit exposure should be expressed. Consider an FX position, for example a cross-currency swap. A natural question is whether risk measures should be computed in the currency cashflows are paid or in the currency they are received. There are good reasons for both choices. If we represent the risk in the receive currency, we implicitly say that the maximum amount we can lose in case of a default of the counterparty is the sum of the receivable cashflows. If the currency of choice is the pay currency we have to accept the fact that, at least in theory, the potential loss is unlimited.

In our view the choice of the reference currency should be adapted to the usage of counterparty exposure. If our goal is simply to report exposure and control it against predefined limits, the correct choice is probably the reporting currency, which could differ from the jurisdiction or the country where the financial institution operates. We could imagine for example that USD could be the currency of choice for a US firm, while EUR seems more natural for European companies. In this case an additional problem, which need to be taken into account, is the FX volatility of the reference currency, which would be incorporated in the computation.

It is may be easier to choose the reference currency when business decisions are involved. If a counterparty is collateralised, it seems natural for example to use as reference the currency of the collateral. Clearly a choice has to be taken when collateral can be posted in different currencies.

Similarly if the counterparty risk is hedged via credit default swaps, it seems natural to use as reference the currency of the hedges, or the currency of the domicile of the counterparty.

12.3.2 Netting and No-Netting Agreements

Aggregation can be performed in different ways, depending on the legal agreements reached with the counterparty, and thus depending on the jurisdictions where the counterparty is located and the transactions performed. In general legal aspects of transactions can be complicated and they are beyond the scope of this book.[7] We will concentrate here mainly on computational aspects.[8]

---

[7]ISDA (International Swap Dealer Association) provides standard templates for contracts between counterparties, known as Master Agreements. The legal agreement for the collateralisation process and for other credit enhancement measures, is described in the so called Credit Support Annex (CSA). These are templates and the exact terms may differ dependending on the wording in the contract.

[8]Even the definition of default is subject to legal interpretation. An interesting example is the bail out of Freddie Mac and Fannie Mae by the US government in September 2008. The two firms were

There are two most common ways of aggregating exposures, one being by netting exposures, i.e. adding algebraically positive and negative values, and the other by no-netting, i.e. adding only positive values and flooring to zero the negative ones.

Consider as an example two call options with different strikes. If aggregation is performed on the base of a netting agreement, the PFE of the portfolio will correspond to only one scenario, i.e. when either the stock is high in value or low. From a qualitative point of view we can imagine that the PFE of the portfolio is equivalent of the PFE of only one instrument. If one of the two positions is short there is also an offsetting effect between the two instruments. On the other hand, if there is no-netting agreement between counterparties, this offsetting effect will not be available as negative exposures will be set equal to zero.

A simulation approach to compute exposure leads naturally to a framework where, at least in principle, aggregation can be performed by adding together the values of the transactions, time step by time step, and path by path. As we have already mentioned, the main challenge is to have the same set of underlying scenarios driving the transaction price distributions, to preserve scenario consistency across the portfolio. Only in very specific cases can quantiles be aggregated directly to obtain portfolio exposure, without having to consider the full price distribution. This means, that only in very specific cases, it is possible to compute the PFE of the trades in isolation, add them together, and still have meaningful values of counterparty exposure.

12.3.3 Break Clauses

A break clause allows either party to terminate a trade at a given time in the future, by settling the mark-to-market value of the trade.[9]

In Fig. 12.1 we consider the effect of a break clause on a portfolio of three swaps with and without break clauses.

Break clauses can be mutual, i.e. both counterparties have the right to break the transaction, or one-sided, when only one counterparty has this right. Break clauses can also be conditional on certain events, (e.g. the mark-to-market value of the trade exceeds a given threshold, the counterparty credit rating falls below a certain level or the counterparty experienced a merger or restructuring) or can be mandatory.

---

placed in a so called 'conservatorship'. This agreement triggered CDS payments. In some cases, e.g. when a sovereign counterparty is involved, it is the risk that the counterparty decides not to honor part of its contracts, that can be relevant. This is sometimes referred as 'repudiation risk'.

[9]Often in practice the terms cancellability and break clause are used interchangeably. We consider as cancellable a trade which has embedded the option to close the deal typically at zero (or at a predefined) value. A break clause, on the other hand involves a cashflow exchange determined by the value of the trade at that time in the given scenario. In this sense a break clause has no economical value, even if it can reduce the counterparty exposure by shortening the period of exposure.

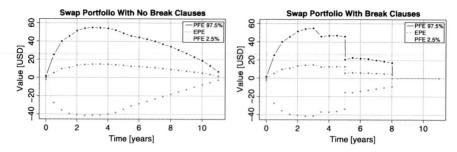

Fig. 12.1 Exposure of a portfolio consisting of three USD swaps, with and without break clauses. The break clauses are at year 3, 5, and 8

From a practical perspective it is often difficult to exercise break clauses as this could damage client relationships. Therefore it is not always clear if one should take them into account when computing counterparty exposure. This is relevant not only for controlling risk, but also and maybe more importantly, for managing credit charges (CVA). As we will describe in more details in Chap. 14 from a pricing perspective it is necessary to include in the value of the transaction the price of credit risk. If a break clause is contemplated in the contract, sometimes the credit valuation is charged only till the date of the potential break. This can create an unbalance if the break is not exercised.[10]

12.4 Collateral Agreements

In general, once aggregated at counterparty level, the values of exposures are checked against predefined limits to ensure that the risk appetite of the firm is not exceeded. In many cases, especially when the credit quality of the counterparty is considered at risk, measures are taken to mitigate exposure. A common device is to agree with the counterparty to have collateral posted in a segregated account,[11] to offset claims in case of default of one of the counterparties. Collateral can be in form of cash or securities and, therefore, needs to be modelled similarly to the underlying OTC portfolio, in order to compute potential future values.

In addition collateral agreements have specific features, which need to be taken into account.[12] Typical examples are the following.

---

[10]These considerations have in reality deeper consequences and involve valuation practice and standards. One question that needs to be answered is how a counterparty should value a transaction, i.e. whether it should include the credit valuation adjustment (CVA) as well as debt valuation adjustment (DVA) and what should be the claim in case of default.

[11]How counterparties can use this collateral (rehypotecation) is also subject to extensive discussions.

[12]Collateral agreements and treatment of collateral in case of default of the counterparty is an art in itself. Considerations which seem to be obvious from a commercial perspective can become

(i) *Initial Margin*. This is the minimum value of collateral that the counterparty has to post regardless of the value of the exposure. This is also the initial amount that the counterparty posts when it enters into the CSA agreement.

(ii) *Margin Calls*. In general counterparties agree to have the possibility of changing the amount of collateral held to reflect changes in value of the transacted portfolio. In addition collateral agreements can be one way or two ways. If the agreement is one way, then any collateral posted will not be returned to the counterparty until the end of the agreement.

(iii) *Margin Dates*. Once margin calls have been agreed it is necessary to also define when they can occur. Often calls are performed daily. In some cases, however, the contract can specify their frequency. Weekly, monthly or quarterly calls can be a common feature of the contract.

(iv) *Threshold*. Often counterparties decide to tolerate a certain amount of exposure regardless of other market or portfolio conditions before starting calling for collateral. This amount is the threshold.

(v) *Minimum Transfer Amount (MTA)*. As the value of the portfolio changes, to avoid relatively small payments of collateral, it is often convenient to define a minimum amount that will be transferred and below which counterparties will not perform a margin call. At a payment date and when a margin call is due, if the collateralised exposure is above the threshold and the excess is above the MTA, then a payment is due.

12.4.1 Counterparty Exposure of Collateralised Counterparties

Indicate with V_t the value at time t of the derivative portfolio transacted between two counterparties and with $C_t = \Phi_t B_t$ the value of collateral; Φ_t is the number of collateral units, B_t the value of each unit, and T_1 and T_2 are the possibly different thresholds between counterparties. To compute counterparty exposure of a collateralised portfolio we need to,

(i) Compute the price distribution V_t of the derivative portfolio taking into account netting agreements and break clauses as indicated in the previous sections.

(ii) Compute the price distribution of the collateral C_t. If the collateral is cash this computation is relatively simple. If, however, the collateral held are notes, its computation can be as sophisticated as the computation of the traded portfolio.[13]

(iii) Determine the unit of collateral exchanged at each margin call date, according to the features of the collateral agreement, the value of V_t and the value of C_t.

---

a source of litigation in court. An example is the netting of collateral with the OTC positions, which we have just described. In some cases, while the defaulted company holds the collateral, its counterparty still has to pay in full its obligation without having the possibility of netting it by the amount of collateral posted.

[13] Sometimes structured notes or stocks can be used as collateral.

The amount of collateral exchanged at each margin call date t_i can be written as follows.

$$\Delta_{collateral} = (V_{t_i} - T_1)^+ - (-V_{t_i} - T_2)^+ - \Phi_{t_{i-1}} B_{t_i}. \qquad (12.6)$$

Negative values of collateral mean that collateral is released by the counterparty. The initial collateral holding corresponds to the initial margin. Note also that at each call date t_i the number of unit of collateral held has been determined at time t_{i-1}, but, as we have to take into account the stochastic behaviour of the collateral, the valuation is performed at time t_i. If the collateral agreement is one way, i.e. if only one counterpart posts collateral, then the margin call is given by

$$\Delta Collateral_{1 way} = (\Delta_{collateral})^+. \qquad (12.7)$$

If a Minimum Transfer Amount has been agreed

$$\Delta_{collateral} = \Delta_{collateral} \mathbb{1}_{\Delta_{collateral} > MTA}. \qquad (12.8)$$

Often in practice the collateral is assumed to be cash and, thus, can be modelled simply as an FX process. If the collateral is a note it can be convenient to consider it as a cash equivalent, possibly reducing its value by a certain amount, which takes into account its potential moves. This amount is often called *haircut*.

12.4.2 Examples

To describe the basic mechanics of the collateralisation process, consider the following stylised example. Assume to have agreed with a counterparty a two way CSA with threshold of 1.5 mUSD, minimum transfer amount of 100 kUSD, daily margin calls. As the market moves the process evolves as follows (see also Table 12.1 below).

– *Day 1.* The counterparty enters into the collateral agreement and as the value of the transaction is at par (and thus below the threshold) no transfer of cash is necessary.
– *Day 2.* The trade moves against the counterparty by 1 mUSD, which is still below the threshold of 1.5 mUSD, so no transfer is necessary.
– *Day 3.* The trade moves further against the counterparty and the collateralised exposure is now 1.55 mUSD which is above the threshold of 1.5 mUSD. As the transfer amount is below the MTA, no margin call is made.
– *Days 4, 5, 6.* Day 4 the trade value increases to 2.9 mUSD. As the move is larger than the MTA, 1.4 mUSD collateral is transferred. Additional collateral is called and then released in Day 5 and 6, respectively, as the move is greater then the minimum transfer amount.
– *Days 7, 8.* No collateral is transferred as the move of the trade from the last collateral call in day 6 is within the MTA.

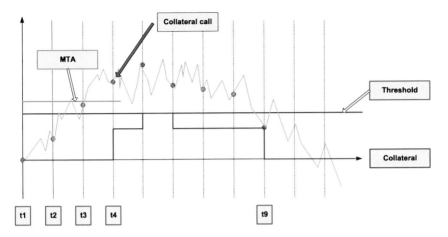

Fig. 12.2 Example of collateralised transaction with daily margin calls, threshold 1.5 mUSD, MTA 100 kUSD

Table 12.1 Example of collateralised transaction. The threshold is 1.5 mUSD, the minimum transfer amount 100 kUSD, and margin calls are two ways, performed daily. Positive collateral means that we held it. Positive portfolio or transaction means that the value is in the money for the counterparty

	Collateral	Transaction Value	Portfolio Value	Transfer Amount
Day 1	0.00	0.00	0.00	0.00
Day 2	0.00	1.00	1.00	0.00
Day 3	0.00	1.55	0.00	0.00
Day 4	1.40	2.90	1.50	1.40
Day 5	2.00	3.50	1.50	0.60
Day 6	1.30	2.80	1.50	−0.70
Day 7	1.30	2.75	1.50	0.00
Day 8	1.30	2.70	−1.00	0.00
Day 9	0.00	1.40	0.05	−1.30

– *Day 9*. The value of the trade further reduces going below the threshold. All collateral is released to the counterparty.

As a further example consider now the USD swap portfolio we analysed in Sect. 12.3.3, where we discussed break clauses, and assume we have a CSA in place. We want to compute exposure taking into account collateral. In the first panel of Fig. 12.3 we consider semi-annual margin calls performed only one way. In other words collateral is not released when the market moves in favour (from a credit perspective) of the counterparty. For this reason, while the 97.5% PFE is capped, the 2.5% PFE is not affected by collateral. We have also assumed a threshold of 10 m USD. This is reflected by the fact that the EPE, being smaller that the threshold, is not affected by collateral calls. In the second panel, we consider the same portfolio, but assume two way margin calls, i.e. the collateral is released to the counterparty when it makes a market profit (and thus have counterparty risk to us).

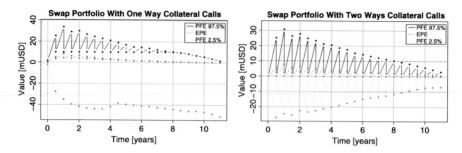

Fig. 12.3 Exposure of a portfolio of three USD swaps with one way (first panel) and two ways (second panel) collateral agreement. Margins are called every six months. The one way collateral has also a threshold of 10 m USD. Collateral is assumed to be cash in the same currency of the swaps (computation performed Jan 2009)

As we mentioned margin calls are often performed daily. We show their effect on the same swap portfolio in the first panel of Fig. 12.4, assuming again threshold of 10 m. In the last panel we have taken into account a CSA with initial margin 10 m, one way collateral, and weekly calls.

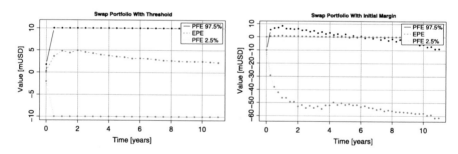

Fig. 12.4 Exposure of a portfolio of three USD swaps. In the first panel we show exposure when a two ways daily margin calls and threshold of 10 m USD collateral agreement has been agreed. In the second panel an initial margin of 10 m USD, and one way margin calls is agreed. As we have plotted the graphs with 3-months frequency, the effect of the daily margin calls is smoothed out. Collateral is assumed to be cash in the same currency as the swaps (computation performed Jan 2009)

12.5 Close-Out Risk

What we have described in the previous sections is simply the value of the portfolio at time t taking into account collateral as an additional instrument in the portfolio. We have focussed on describing the mechanics of the collateral agreement.

The default notification, as well as the liquidation of the portfolio, can be, however, a lengthy process. For these reasons when dealing with collateralised transactions, it is necessary to compute not only the value of the portfolio and of the collateral, taking into account the CSA features. It is also crucial to assess how much, upon default of the counterparty, the traded portfolio could change in value till it is liquidated. Valuing the portfolio changes during the liquidation period involves computing the so called *close-out risk*.

The computation of the close-out risk should be done assuming that the default of the counterparty could happen at any time during the life of the portfolio. This can be achieved by computing the portfolio and collateral price distribution, V_t and C_t, and then by taking into account the fact that during the close-out period δ the value of the portfolio could change, while the amount of collateral remains unchanged,[14]

$$\text{Close-out}_t = (V_{t+\delta} - \Phi_t B_{t+\delta})^+ - (V_t - \Phi_t B_t)^+. \tag{12.9}$$

By doing this computation path-by-path as a post-process operation we obtain a distribution of prices over time and we can take statistical measures to characterise it. Often in practice, the risk measure used to compute close-out is again a high confidence level quantile, such as for example 97.5% or 99% of the price distribution. Credit valuation adjustments (see Chap. 14) need also to take close-out risk into account. In this case the measure to use is the EPE. To assess the length of the liquidation period it is necessary to consider the time needed to realise that the counterparty has defaulted, and the liquidity of the transactions which need to be settled. It can therefore vary from one day, e.g. for swaps in one of the major currencies, to several months, when for example structured products, such as illiquid credit derivatives are traded. It depends also on the size of the position, as selling large volumes of e.g. stocks could affect the market. In general the close-out period takes also into account a so called *grace* period, in which the solvency of the counterparty is checked.

The value of the close-out during the life of the transaction is, at least in principle, the amount of collateral a counterparty should post. This is often called initial margin or haircut.

12.6 Risk Allocation

Once exposures have been aggregated at portfolio level and the portfolio PFE or EPE has been computed, it can be interesting to allocate exposures back to each single transaction, taking into account the portfolio effect. This is useful to assess, for example, which transaction in a portfolio is risk reducing, or to allocate transactions to different business lines. As quantiles are non-additive there is no obvious methodology to compute individual contributions to the overall exposure. We will

---

[14]As we suppose that the counterparty has defaulted we cannot assume to get additional collateral.

consider in the following two possibilities, a naive allocation often used in the financial industry for its simplicity, and the Euler allocation, which has the advantage of taking into account portfolio netting effects.

12.6.1 Naive Allocation

The simplest way to allocate exposure back to trades in a portfolio is by assigning an exposure fraction proportional to the exposure of single trades.

$$r_i = \left(\frac{q_i}{\sum_j q_j} \right) q, \tag{12.10}$$

where q is the PFE or EPE exposure at counterparty level, and q_i is the PFE (or EPE) of each trade considered individually. This approach suffers from the problem that no netting effect is really taken into account. Consider for example the case of two swaps of different maturity, one payer and one receiver. As we can see in Table 12.2 both swaps will have a positive contribution. In this sense a possible alternative is the Euler allocation algorithm [78], defined as the first derivative of the risk measure with respect to the risk factors involved. As we will see in the next section, it can be written as expected value of the risk factor conditional on the portfolio quantile (or shortfall).

12.6.2 Euler Allocation

Suppose we have a portfolio counterparty exposure made of multiple trades X_i, so that $V = \sum_i \lambda_i X_i$, and assume that ρ is a risk measure function. The Euler allocation rule simply states that in this case

$$\rho(V) = \sum_i \lambda_i \frac{\partial \rho(V)}{\partial \lambda_i} = \sum_i r_i, \tag{12.11}$$

where r_i is the risk contribution of trade i to the overall portfolio level (see also Tasche [102]). For our purpose, set $\lambda = 1$ and assume the risk measure ρ to be a quantile, then the following decomposition holds

$$r_i = \frac{\partial q_\alpha}{\partial \lambda_i} = \mathbb{E}(X_i | V = q_\alpha). \tag{12.12}$$

This considerably simplifies the calculation of the contribution, since it turns the problem of calculating a derivative on a Monte Carlo simulation into one of finding the expected value of the individual trades conditional on the portfolio value. Note

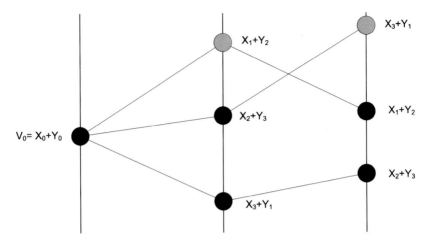

Fig. 12.5 The portfolio consists of two trade X and Y. Suppose we only have three paths, and that the quantile measure is given by taking the maximum point. At the first time step the contributions are $r_x = X_1$ and $r_y = Y_2$. At the second time step the contributions are X_3 and Y_1, respectively

that for simplicity, we have dropped the time subscript t, since this value can be calculated independently at each time slice.

Equation (12.12) shows how to compute contributions by expectation. The computation could be performed by brute force, simply repeating the Monte Carlo simulation and taking then the average. This becomes quickly unfeasible. On the other extreme one could perform only one Monte Carlo simulation and compute the average using only one sample. This is illustrated in Fig. 12.5, where we assume to have a portfolio with two trades.

This method is crude and in many cases will not give a good estimate of the contributions. In particular it can be very sensitive to new simulation runs.

An other possibility to estimate the expectation is to take a bundle of similar paths, and get some sort of expected contribution. The diagram below illustrates this idea.

This method has clear advantages in terms of computational time over the alternative of re-running the simulation. We can improve this method by weighting the samples in order to improve the accuracy of the quantile computation (see also Harrel & Davis [56]).

12.6.3 Comparison with Naive Allocation

Consider as an example a portfolio of two swaps, a three year receiver and a seven year payer swap. As we have seen previously, the simple allocation rule would assign a positive proportion to both swaps, even if there is a netting effect. The Euler allocation would provide a positive contribution to the seven year swap and negative

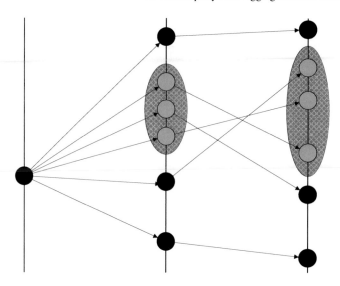

Fig. 12.6 The portfolio consist of two trades. Expectation is computed over a bundle of values which are close

contribution to the three year swap reflecting the netting effect of the two trades. The table below shows the comparative difference between the two contributions.

Table 12.2 Contributions in a portfolio of two swaps, one payer and one receiver. The naive algorithm allocate positive contribution to both trades, while the Euler scheme gives a positive and negative, reflecting netting effect

Year	Naive 3Y	Naive 7Y	Euler 3Y	Euler 7Y	Total
1	1.50	3.94	−3.43	8.88	5.44
2	2.42	5.60	−4.74	12.76	8.02
3	2.12	7.85	−2.82	12.79	9.97
4	0.00	11.56	0.00	11.56	11.56
5	0.00	9.27	0.00	9.27	9.27
6	0.00	6.98	0.00	6.98	6.98
7	0.00	3.60	0.00	3.60	3.60

The advantage of the Euler allocation rule is now clear. The three years swap has a reduction effect on the exposures on the 7 year swap. It reduces in fact the overall maximum exposure from 12.79 in year three to 11.56 in year four. Clearly this reduction effect vanishes after the maturity of the three years swap.

12.6.4 Contribution Calculation of Collateralised Transactions

In principle allocation of counterparty exposure to individual transactions can be performed as described in Sect. 12.6, by treating collateral as an other leg of the portfolio. In practice, however, this can give unstable numerical results. It is therefore more convenient to first compute contributions ignoring collateral and then compute the actual exposure taking into account collateral. The contribution due to the collateral agreement can then be derived as difference between exposure with and without collateral.

Chapter 13
Combining Market and Credit Risk

The valuation approach detailed in Chap. 4 is centered on estimating the distribution of future values of a transaction after having simulated trajectories of the underlying stochastic drivers. When markets are complete, the pricing-by-arbitrage paradigm allows us to price stochastic payoffs as an expectation in a particular measure, namely the one under which the prices of assets are martingales when expressed in units of a chosen numeraire.

In the context of Chap. 4, this means performing the AMC estimation algorithm on simulations drawn from this martingale measure, which we have denoted throughout by $\mathbb{N}$. In practice, one needs to know the probability distribution of future trade values also in measures other than that under which the simulation takes place. For example:

(i) Asset price processes evolve in the real-world measure, and not the risk-neutral one used for pricing.

(ii) In assessing market risk, one would look at the value distribution under a measure where chosen underlying risk factors are conditioned to have a chosen behaviour, e.g. under a stress scenario.

(iii) In measuring counterparty credit risk, issues of right-way/wrong-way risk arise. This happens when the value of a transaction and the quality of the counterparty are not independent. In this case, conditioning on the event of counterparty default alters the value distribution, so that credit exposure should be analysed in a measure different to that used for pricing.

All the above problems reduce to analysing the value distribution in a probability measure different to that used for simulation. Importantly, we would like such analysis to be possible *without* having to perform additional simulations, and we show in this chapter how this can be achieved using well-known change-of-measure techniques. In particular, we present and compare two approaches to right-way/wrong-way risk issues, a topic of central importance from a regulatory perspective.[1]

---

[1] Under Basel II right-way/wrong-way exposures have to be identified and computed accordingly.

G. Cesari et al., *Modelling, Pricing, and Hedging Counterparty Credit Exposure,*
Springer Finance, DOI 10.1007/978-3-642-04454-0_13,
© Springer-Verlag Berlin Heidelberg 2009

13.1 Change of Measure: Practical Implementation

In Appendix B we describe the classical mathematical set-up and notation to change measure. In this section we highlight how these techniques can be applied in practice within our framework.

Our AMC estimation algorithm results in samples of values of the distribution of future values of the transaction under the measure $\mathbb{N}$, with, say, $\hat{F}_t^{\mathbb{N}}$, being the time-t empirical distribution under $\mathbb{N}$, obtained as

$$\hat{F}_t^{\mathbb{N}}(v) = \hat{\mathbb{E}}\left[\mathbb{1}_{V_t < v}\right] = n^{-1} \sum_{i=1}^{n} \mathbb{1}_{V_t^{(i)} < v}, \quad v \in \mathbb{R}; \tag{13.1}$$

here, V_t is the distribution of time-t prices (which has been obtained in $\mathbb{N}$). The ^ indicates quantities obtained empirically from sample distributions.

The goal is now to use the change of measure technique to obtain $\hat{F}_t^{\mathbb{P}}$, the empirical distribution under an alternative measure $\mathbb{P}$. For this purpose we can use the Radon-Nikodym derivative

$$\zeta_t = \frac{d\mathbb{P}}{d\mathbb{N}}\bigg|_{\mathscr{F}_t} \tag{13.2}$$

to write, for each $v \in \mathbb{R}$,

$$F_t^{\mathbb{P}}(v) := \mathbb{P}[V_t \leq v]$$

$$= \mathbb{E}^{\mathbb{P}}[\mathbb{1}\{V_t \leq v\}]$$

$$= \mathbb{E}[\mathbb{1}\{V_t \leq v\}\zeta_t]. \tag{13.3}$$

Of course, expression (13.3) will need to be computed empirically from the sample of values $\{V_t^{(i)}\}$ and the corresponding values $\{\zeta_t^{(i)}\}$, resulting in the empirical version, $\hat{F}_t^{\mathbb{P}}$, of $F_t^{\mathbb{P}}$, by replacing expectations with sample means.

Using the empirical estimates of the CDF's $\hat{F}_t^{\mathbb{P}}$ and $\hat{F}_t^{\mathbb{N}}$, we obtain a sample of V_t under $\mathbb{P}$ by setting[2]

$$\tilde{V}_t^{(i)} = \left(\hat{F}_t^{\mathbb{P}}\right)^{-1} \left(\hat{F}_t^{\mathbb{N}}\left(V_t^{(i)}\right)\right), \tag{13.4}$$

where the superscript (i) indicates the i'th of a sample of values, and where the ~ signifies values of V_t under the changed measure $\mathbb{P}$. From the sample $\{\tilde{V}_t^{(i)}\}$, risk measures such empirical quantiles under the measure $\mathbb{P}$ can now be computed in the same way as for the original sample of values $\{V_t^{(i)}\}$.

To illustrate with a particular example, consider having a transaction driven by d risk factors, and suppose we want to compute exposure in the real-world measure. Let $\mathbf{X}$ be standard Brownian Motion in $\mathbb{R}^d$, and let $\mathbf{Y} \equiv \mathbf{R}\mathbf{X}$ (with $\mathbf{R}^\mathsf{T}\mathbf{R}$ positive

---

[2] ... using, for example, an interpolation procedure to invert $\hat{F}_t^{\mathbb{P}}$ ....

semi-definite) represent the risk factor evolution. It can be shown that taking the Radon-Nikodym derivative ζ in (B.4) with

$$\gamma_t = (\mathbf{R}^\top \mathbf{R})^{-1}\mathbf{R}^\top \delta, \quad \delta \in \mathbb{R}^d, \tag{13.5}$$

and δ constant, imparts to Y a drift of δ. In order to change from the martingale measure $\mathbb{N}$ to the real-world measure $\mathbb{P}$, the i'th element of δ should be set equal to the market price of risk[3] pertaining to the i'th risk factor.

In what follows, we turn to practical situations in which the above can be used.

13.2 Exposure under Real-World Measure

The exercise of estimating credit exposure under the real-world measure becomes especially important in market conditions where trade underlying processes exhibit a marked drift. Consider, for example, the price of oil in July 2007. The forward price for oil delivered in one year from this date, was approximately $72. The max PFE 97.5% exposure for a transaction to buy 100 barrels forward at this price would have been estimated at around $5000. The actual price of oil at expiration of this forward (July 2008) was in the region of $130, which means that the *realised* payoff on the forward at expiration would be closer to $6000, exceeding the PFE that would have been estimated at trade inception.

In a transaction such as this, it would have been helpful to analyse, at trade inception, the change in exposure from imposing a drift on the underlying risk driver. In this case, the realised exposure of $6000 would correspond to a drift change $\delta = 0.3$ in expression (13.5). In the log-normal case, and assuming a volatility of 30%, this translates to the risk driver (oil price) having a return of 9% in excess of the risk-free rate.

We show in Fig. 13.1 the exposure of an oil forward computed with market data of September 2007.

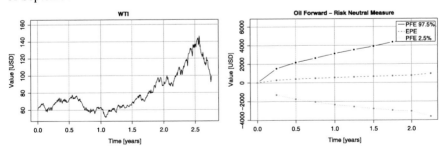

Fig. 13.1 Time series of WTI from January 2006 to September 2008 and exposure of a forward on WTI computed with market data of September 2007

---

[3]For a log-normally distributed risk factor, this is the ratio of excess return to volatility, otherwise known as the Sharpe ratio.

The left panel shows a time series of WTI; estimating a drift from such time series can be very difficult, not only for computational reasons, but also because time series follow different regimes which can be drastically different. For this reason, the particular drift δ chosen for examining the price distribution should be thought of more as a possible scenario than as a definite estimated value. Despite this, such analysis can prove useful to judge how prices will behave in given regimes.

13.3 Stress Testing

At a general level, stress testing involves investigating the behaviour of the price distribution of a position contingent on chosen scenarios of the underlying risk drivers. For example, in the case of a portfolio sensitive to GBP interest rates, one would be interested in what happens to the position if rates were to move by a chosen amount or to reach a certain level.

In order to perform such an investigation, the appropriate change of measure would be to impose the desired drift on the risk driver of interest. In more detail, suppose we give ourselves a set of different measures

$$\mathbb{P}_1, \ldots, \mathbb{P}_n \tag{13.6}$$

corresponding to regimes in which the risk driver of interest has drifts

$$\delta_1, \ldots, \delta_n. \tag{13.7}$$

The interpretation of the $\mathbb{P}_i$ is as different 'stress' scenarios in which the growth of the risk factor would be accelerated or slowed down. Given such scenarios, it is natural to look for a summary measure of the risk posed by holding products sensitive to the risk driver.

One such summary measure turns out in fact to have the desirable property referred to as *coherence* (see also Chap. 12). It is a theorem (see for example [78]) that if $\mathscr{P} = \{\mathbb{P}_i\}$ is a set of probability measures, the risk measure ρ defined by

$$\rho(V) := \sup_{\mathbb{P}_i \in \mathscr{P}} \mathbb{E}^{\mathbb{P}_i}[V] \tag{13.8}$$

on the set

$$L(\mathscr{P}) := \{V : \mathbb{E}^{\mathbb{P}_i}[V] < \infty \text{ for all } \mathbb{P}_i \in \mathscr{P}\} \tag{13.9}$$

is a coherent risk measure on $L(\mathscr{P})$. In particular, such a measure satisfies the property of subadditivity.

Computing the risk measure ρ, therefore, amounts to computing the distribution mean under a set of different measures, and then taking the largest amongst those means. To take a specific example, consider a scenario in which the price of gold undergoes unprecedented increase;[4] this corresponds to a large positive drift γ, corresponding to the measure $\mathbb{P}^*$ say. If V_t is the price distribution at t of a trade that has gold as its underlying driver, then

$$\rho(V_t) = \max\{\mathbb{E}^*(V_t), \mathbb{E}(V_t)\} \tag{13.10}$$

is the coherent risk measure in this case. If V is the price of a portfolio of transactions, using a coherent risk measure to gauge the riskiness of V is of crucial importance (again, see [78]).

By way of illustration, we show in Fig. 13.2, for a trade sensitive to oil prices, the extreme quantiles and means of the price distributions under the pricing measure $\mathbb{N}$ and an alternative measure where oil appreciates at a higher rate. At each time, the larger of the two means corresponds to ρ.

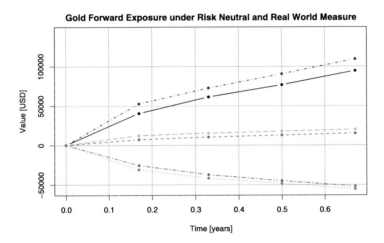

Fig. 13.2 Extreme quantiles and mean of a gold forward transaction, estimated under the pricing measure as well as under a 'stressed' measure where the gold price appreciates at a higher rate. The larger of the two means results in a coherent risk measure for gauging the transaction

13.4 Right-Way/Wrong-Way Exposure

By our definition, and in line with business practice, the credit exposure faced by a party in a transaction is the amount that party would lose in the event of the counterparty defaulting. This means that the amount exposed at any given time t is the distribution of the value of the transaction, *conditional on* the counterparty defaulting at t. Very often, this conditioning does not affect at all (or does so only to a small extent) the price distribution of the transaction—in these cases, credit expo-

---

[4]Gold became 9% more expensive over the course of a single day, September 17, 2008.

sure is equivalent to price exposure, so that estimating price distributions is all that is needed.

Right-Way Exposure (and Wrong-Way Exposure) refers to the case when the distribution of future values of the transaction is *not* independent of the credit worthiness of the counterparty. In other words, conditioning on counterparty default *does* affect the price distribution of the transaction and one then needs to find ways of taking the conditioning into account. The exposure is called right way (resp. wrong way) if the value of the transaction is negatively (resp. positively) correlated to the counterparty credit quality.

To take an extreme example, suppose one were to buy from a counterparty a call option written on that counterparty's stock. Clearly, the relevance of counterparty default to the stock option price cannot be dismissed; in fact at default we can expect the stock (and hence the call option and the credit exposure on it) to be worth nothing. Analogously, the counterparty exposure on a put option would be equal to the option notional.

In general the link between quality of the counterparty and exposure of the trade is not as straightforward as in the previous examples. For instance, consider a commodity trade that depends on values of oil and gas. Suppose the counterparty in the trade is a firm whose business is related to these commodities, so that default of the firm would be likely if the prices of the commodities drop significantly. In this case, the event of counterparty default is consistent with lower gas and oil prices. To assess the credit exposure arising from the transaction (which involves conditioning on counterparty default) one should therefore look at the value distribution in a measure where the underlying oil and gas prices have a large negative drift.

There are several ways in which this intuitive idea can be put into practice. One way is the change of measure technique discussed above. Another, which we describe below, stems from a Merton-like model for the event of counterparty. We will see that this method can be used only when the payoff of the transaction is monotonic in the risk factors. We close by describing a few illustrative examples, comparing results from the two approaches outlined.

13.4.1 Right-Way/Wrong-Way Exposure: Merton Approach

The well-known Merton model [80] for default links the default of a firm to the difference between assets of a firm and its liabilities. At any time, then, this distance-to-default is indicative of the propensity of that firm to default. The application of this idea to right-way/wrong-way risk is by allowing the distance-to-default process for the counterparty to have covariance with the process underlying the transaction entered into.

To this end, we choose a fixed time horizon and suppose the economy at that time to be driven by some market factor[5] $Z \sim N(0, 1)$. In general, both the transaction

---

[5]We think of Z being univariate. Extending to multiple market factors could make calibration more intricate, but the mathematics not much more so.

value and the likelihood of counterparty default depend on the market factor Z. To reflect this, we define a random variable X,

$$X = \rho Z + \bar{\rho} \tilde{X} \tag{13.11}$$

where $\tilde{X} \sim N(0, 1)$ is independent of Z and where $\rho \in [-1, 1]$, $\bar{\rho} = \sqrt{1 - \rho^2}$. We now use X to model the default of the counterparty as

$$\text{Counterparty defaults} \Leftrightarrow \{X < C\}, \tag{13.12}$$

which, because we have modelled X as $N(0, 1)$, immediately implies that C must satisfy

$$p = \Phi(C), \tag{13.13}$$

where p is the probability of counterparty default and Φ the cumulative standard normal distribution. Note that this is a static model with all quantities C, p, and X relating to a particular time point of interest.

Mathematically, (13.12) and (13.13) allow us to relate default to a probability distribution that is well known to us, even if the random variate X is a priori unobservable. However, if we think of X as being the (logarithm of) returns of the counterparty firm value process, then (13.12) says that default happens when returns are lower than a certain value, and this is intuitive. In summary, the unobserved variable X, introduced as a mathematical device to model the default event, has an interpretation as the return of counterparty firm value, possibly after a transformation and/or re-scaling.

We would like a model that acknowledges the fact that the product price distribution is altered given the event of counterparty default. That is, we are interested in knowing the distribution of the market factor Z conditional on the information that the counterparty has defaulted. Now, from (13.11) we obtain[6] this as

$$\mathbb{P}[Z = z \mid X < C] = \mathbb{P}[X < C \mid Z = z] \frac{\mathbb{P}[Z = z]}{\mathbb{P}[X < C]}$$

$$= \frac{\Phi\left(\frac{C - \rho z}{\bar{\rho}}\right)}{\Phi(C)} \phi(z) dz$$

$$=: \tilde{\phi}(z) dz, \tag{13.14}$$

in comparison with the unconditional probability

$$\mathbb{P}[Z = z] = \phi(z), \tag{13.15}$$

ϕ being the standard normal density function.[7]

---

[6] We abuse notation by writing $Z = z$ to mean $Z \in [z, z + dz)$.

[7] We use $\mathbb{P}$ to highlight that we are not necessarily under the risk neutral measure.

We note that the effect of a non-zero value for ρ is to weight, for each z, the original density $\phi(z)$ according to whether the occurrence of default makes observing the value z more or less likely. In effect, $\tilde{\phi}$ is the density under a new measure (implied by conditioning on the default event) of the random variable Z (whose unconditioned distribution is standard normal). In this case, the Radon-Nikodym derivative of $\tilde{\phi}$ with respect to ϕ is given by

$$\zeta \equiv \zeta(Z) = \Phi\left(\frac{C - \rho Z}{\bar{\rho}}\right)/\Phi(C). \tag{13.16}$$

Figure 13.3 shows the effect of changing the dependence parameter ρ on the shape of the conditional distribution of the driving market factor Z in (13.11). For positive ρ, counterparty default is associated with low values of X and hence of Z, giving rise to the curves appearing on the left side of the plot. Conversely, a negative ρ causes a shift to the right of the mass of the unconditional distribution.

Distribution of Z conditional on counterparty default

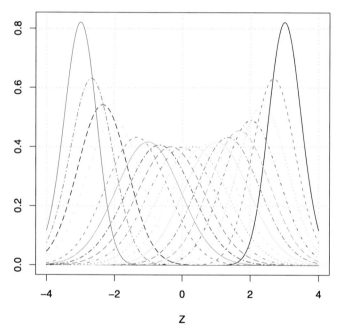

Fig. 13.3 Distribution of the Gaussian factor Z conditional on the event of counterparty default, $\{X < C\}$, for values of ρ ranging in $[-1, 1]$. The marginal default probability is $0.001 = \Phi(C)$. The central curve (for $\rho = 0$) coincides with the unconditioned density ϕ

Using the prescription (13.16), it is straightforward to estimate statistics of the value distribution under the conditioned measure. Thus, for example, given a level of confidence q, the q-tile exposure for V under the conditioned measure is the solution $x_q^{(\rho)}$ to

$$
\begin{aligned}
q &= \mathbb{P}\{V \leq x_q^{(\rho)}\} \\
&= \int_{\{z:V \leq x_q^{(\rho)}\}} \tilde{\phi}(z)dz \\
&= \int_{\{z:V \leq x_q^{(\rho)}\}} \phi(z)\frac{\Phi(\frac{C-\rho z}{\bar{\rho}})}{\Phi(C)}dz \\
&= \int_{\{z:V \leq x_q^{(\rho)}\}} \phi(z)\frac{\Phi(\frac{C-\rho z}{\bar{\rho}})}{\Phi(C)}dz \bigg/ \int_{-\infty}^{\infty} \frac{\Phi(\frac{C-\rho z}{\bar{\rho}})}{\Phi(C)}\phi(z)dz \\
&\equiv \mathbb{E}^{\mathbb{P}}\left[\mathbb{1}_{V \leq x_q^{(\rho)}}\frac{\mathbb{P}\{X < C \mid Z\}}{\mathbb{P}\{X < C\}}\right] \bigg/ \mathbb{E}^{\mathbb{P}}\left[\frac{\mathbb{P}\{X < C \mid Z\}}{\mathbb{P}\{X < C\}}\right],
\end{aligned}
\tag{13.17}
$$

where the denominator in the last two expressions is equal to unity by virtue of the fact that the expected conditional default probability equates to the marginal default probability. From (13.17), if we suppose that we have simulated V by first simulating n draws $\{z^{(i)}\}$, $i = 1, 2, \ldots, N$, from the (unconditioned) law ϕ of Z, we can estimate $x_q^{(\rho)}$ by solving for $x_q^{(\rho)}$ in

$$
q = \sum_{\{z:V \leq x_q^{(\rho)}\}} \frac{\Phi(\frac{C-\rho z}{\bar{\rho}})}{\Phi(C)} \bigg/ \sum_{\{z\}} \frac{\Phi(\frac{C-\rho z}{\bar{\rho}})}{\Phi(C)},
\tag{13.18}
$$

where the sum in the denominator is over all simulated values $\{z^{(i)}\}$. Of course, the exact value of the denominator is unity, but in practice we include it in numerical calculations to ensure that the conditioned distribution has the correct total mass. We re-iterate that (13.18) does *not* require simulating the product future value distribution V under the dependency-adjusted measure whose density is $\tilde{\phi}(\cdot)$.

13.4.2 The Inverse Problem

Thus far, we have looked at the problem of obtaining estimates from the value distribution conditioned on default. It is also of practical interest to be able to solve the inverse problem, that is, to find the probability of counterparty default *conditional* on some observed level of exposure. The interest in doing this lies in (i) being able to obtain a view on default probability conditional on a given exposure (and hence, implicitly, conditional on a given scenario of the market variable), and (ii) allowing one to map the (univariate) price distribution dictating exposure levels to a distribution (in our model the univariate normal market variable Z) that partly drives default. Importantly, this gets around the problem created when the product price is a function of several state variables, themselves correlated with the market variable affecting default.

The simple key idea is as follows. Suppose V denotes the price distribution at some fixed time t of interest. What we *know* is that V arises from a complicated pricing function, in general depending on several market factors which are not independent. On the other hand, what we *would like* is to write $V = G(Z)$ as a function G of a (standard normal) random variable Z. This would allow us to find out what value of the market variable Z would correspond to some chosen value of the price distribution V.

In more detail, if we let F denote the empirical distribution of the price distribution V, then it is plain that

$$F(V) \tag{13.19}$$

has a uniform distribution. In turn, recycling the same idea one more time we get our desired standard normal random variable from

$$Z = \Phi^{-1}\Big(F(V)\Big), \tag{13.20}$$

with $\Phi(\cdot)$ being, of course, the standard normal distribution function. Given the event

$$\{V = v\}, \tag{13.21}$$

and hence, from (13.20), that

$$\{Z = z = \Phi^{-1}(F(v))\}, \tag{13.22}$$

we can obtain the conditional counterparty default probability as

$$\mathbb{P}[X < C \mid Z = z] = \mathbb{P}[X < C \mid V = v]$$

$$= \Phi\left(\frac{C - \rho z}{\bar{\rho}}\right)$$

$$= \frac{\tilde{\phi}(z)}{\phi(z)}\Phi(C). \tag{13.23}$$

Note that the idea (13.22) of mapping the distribution V onto a standard normal random variable Z fails as soon as the payoff giving rise to the distribution V is not monotonic in its underlying risk factors.

13.4.3 Example 1: Call Option on Stock

The first example we consider is a call option written on a stock with constant volatility. We estimated the change in exposure that results under the assumption that the event of counterparty default causes the underlying stock price process to drift upwards at the rate of 0.05σ, 0.1σ or 0.15σ in excess of the risk-free rate. We then repeated the calculation using the Merton approach, with a default probability intensity $\lambda = 0.01$ and with values of ρ that give similar results to the change-of-measure approach.

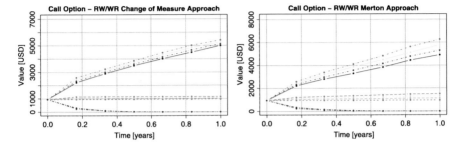

Fig. 13.4 Exposure profiles (97.5%, 2.5% PFE, and EPE) of an at-the-money call option written on a stock. The stock has spot value 100 USD and volatility $\sigma = 20\%$ and the option is of maturity 1 year. The left panel shows result from the change-of-measure approach computed for cases when the stock price process has drift 0.05σ, 0.1σ and 0.15σ in excess of the risk-free rate. Higher drift leads to increasing exposure, since likelihood of scenarios with the call option being in the money is increased. The right panel shows results from the Merton model approach with an intensity of default $\lambda = 1\%$, and with values ρ chosen to give results similar to those from the change-of-measure approach

In Fig. 13.5, using the Merton model, we show the dependence of exposure on the level of correlation. Exposure decreases as correlation between the underlying stock and the quality of the counterparty increases. When the counterparty trades a call option on its own stock exposure becomes zero.

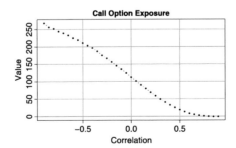

Fig. 13.5 The exposure level for ρ ranging in $[-1, 1]$, with $q = 97.5\%$. The product is an at-the-money call option on a stock with strike 100, volatility 20%, fixed interest rate 5%, and expiry 5 years. The exposure is that at the $t = 2.5$-year point; for the case $\rho = 0$, this of course coincides with the quantile of the distribution of V in the unconditional measure for Z

13.4.4 Example 2: Call Put Structure on Oil

The second example is similar to the first, but has the oil price process as underlying. We consider a structure made up of a short put with strike 95%, a long put with strike 90%, and a long call with strike 105%, with the same maturity. We consider exposure using different drifts. Since the portfolio payoff is monotonic increasing in the price of the underlying, the exposure posed by the option increases with the excess drift.

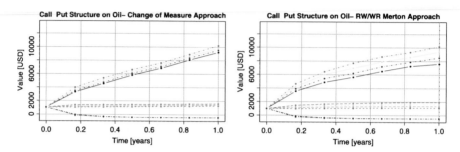

Fig. 13.6 Exposure profiles (97.5%, 2.5% PFE, and EPE) of a portfolio with a long call with strike 105%, a short put with strike 95%, and a long put with strike 90%, all of maturity equal to one year. The option underlying for this example is oil (WTI), with a spot value of 95% and volatility $\sigma = 35\%$. Exposure is computed using a drift of $\gamma = 0.05\sigma, 0.1\sigma, 0.15\sigma$ in excess of the risk-free rate. The figure on the right panel reports estimates of exposure computed following the Merton approach for an intensity of counterparty default $\lambda = 1\%$, and correlation levels $\rho = 5\%, 10\%, 15\%$

13.4.5 Example 3: Cross-Currency Swap on USD-GBP

As a final example, we consider a cross-currency swap where we receive USD and pay GBP. The change of measure we employ in this example is to assume that GBP appreciates against USD. Such a scenario would be relevant, for instance, when the counterparty is a US-based firm with GBP liabilities. In this case, appreciation in GBP against USD would be expected to increase the risk of default of the counterparty.

13.4.6 Comparison with the Change of Measure Approach

The two approaches to right-way/wrong-way risk that we have described are special cases of a change of measure. However, the Gaussian-based model takes as input a correlation parameter between Gaussian drivers of the transaction value and the counterparty distance-to-default indicator, while in the change-of-measure approach the conditioning on counterparty default is expressed by imposing a chosen

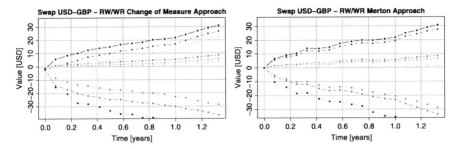

Fig. 13.7 Exposure profiles (97.5%, 2.5% PFE, and EPE) for an FX swap, of maturity 2 years, where we receive USD and pay GBP. The 2-year USD zero rate is approximately 3%, the GBP rate at 5.2%. The exchange rate volatility is $\sigma = 10\%$. For the change of measure, we chose excess drifts $\gamma = 1.0\sigma$, 2.0σ on the USD/GBP exchange rate, that is, regimes where GBP appreciates against USD. Since the transaction receives USD, this causes the exposure level to drop. On the right hand panel we consider the Merton model approach, with $\lambda = 1\%$ and $\rho = 10\%$, 50%, 70%

drift on the relevant risk driver that one believes would be affected by, or would be responsible for, the default event.

In the Merton approach, the linear correlation parameter ρ relates to the dependence between transaction values and couterparty default. As such, it is a difficult quantity to estimate. The method we have commonly used is to estimate the linear correlation between the counterparty equity returns and those of the risk driver underlying the trade. In estimating exposure, it is then appropriate to use a range of values for the correlation parameter, in order to also have an idea of the sensitivity to ρ of the exposure computed.

The change-of-measure approach, on the other hand, allows one to impose a chosen drift on a particular risk driver in the transaction. The advantage of this is that one can explicitly control the evolution of the risk driver of interest.

We will analyse in the next chapter wrong-way/right-way risk in the context of CVA computation, and we will show how, at least in some cases, it can be taken into account during the simulation of the underlying scenarios.

Chapter 14
Pricing Counterparty Credit Risk

We have analysed in the previous chapters the most straightforward ways of mitigating the risk of default of a counterparty, namely by imposing limits on transacted notional amounts and by negotiating collateral agreements with the counterparty.

A more flexible alternative is to buy protection or insurance[1] on the given counterparty, typically in the form of a Credit Default Swap (CDS). In practice, however, risk mitigation via CDSs is not always straightforward.

(i) The CDS market for the counterparty may not be liquid enough, and may not offer instruments of the desired maturity. In practice, when liquidity is lacking, it becomes necessary to buy protection not on the counterparty, but on more liquid credit-linked indices.[2]

(ii) There is no unique way to determine the notional values for the CDS contracts that need to be purchased.

(iii) The recovery value of a defaulted bond delivered under a CDS contract has little to do with the fraction that can be recovered from the contract being hedged.

(iv) Dynamically varying the amount and time profile of protection after inception of the trade may be prohibitively expensive.

In any case, putting in place a static hedge by simply buying CDS protection at trade inception is akin to trying to hedge the risk of a stock option by holding a fixed amount of the underlying stock. The ultimate goal should be to hedge credit exposure dynamically, by actively rebalancing a portfolio of instruments to ensure that its value at any point in time (in particular, at the random time at which the counterparty defaults) matches the value (when positive) of the trade being hedged.

---

[1] We use here the term "insurance" in a generic way. From the legal point of view an insurance has specific characteristics, which distinguish them from protection obtained via CDS.

[2] These indices are available on a wide range of names and products. For example, CDX and iTraxx cover corporate names mainly in North America and Europe, while ABX and CMBX are indices based on ABS (asset backed securities). The main advantage of using indices is liquidity and lower transaction cost. The disadvantage is that they are generic, i.e. not tailored to hedge the specific portfolio at hand, and will not pay the full protection amount upon default of the counterparty. These indices should be used mainly to hedge the market risk coming from CDS spreads.

G. Cesari et al., *Modelling, Pricing, and Hedging Counterparty Credit Exposure,*   215
Springer Finance, DOI 10.1007/978-3-642-04454-0_14,
© Springer-Verlag Berlin Heidelberg 2009

Such a transaction that replicates the credit exposure posed by another transaction, is often referred to as a *contingent credit default swap* (C-CDS). The time-zero value of the C-CDS, which is simply the amount of funds required for the initial set up of a replicating portfolio, is generally called *credit charge*, or *credit valuation adjustment* (CVA).

Note that since CVA is part of an Investment Bank's profit and loss, it becomes a major source of P&L volatility. Therefore there is a need to hedge CVA, not just to protect against counterparties defaulting on their obligations, but also to reduce this volatility.

In what follows, we describe in more detail the calculation of CVA for any type of transaction. We then proceed to show how to replicate a C-CDS using a self-financing portfolio of liquid market instruments whose prices are linked to the processes underlying the transaction being hedged.

14.1 Credit Valuation Adjustment and Static Hedging

The CVA of a given transaction is the fair value of protection that compensates for the loss incurred when the counterparty defaults. This is similar to the value of the protection leg of a standard CDS. However, while a standard CDS pays a fixed amount adjusted by the recovery rate, in this case the payment is linked to the (stochastic) value of an underlying transaction or portfolio of transactions. It is important to emphasise that the CVA indicates how much it would cost to set up a portfolio that replicates the credit exposure being hedged, but *not* how this portfolio should be structured.

In this section we will see how to write the CVA of a transaction in terms of the counterparty credit spread and the modified EPE of the transaction. We will then discuss how to construct the replicating portfolio itself.

Denote by $V \equiv \{V_t\}_{t \geq 0}$ the value of the transaction on which protection is sought, R_V the recovery which we expect to make on the portfolio should the counterparty default, τ the random time of default of the counterparty, and let N be, as usual, the chosen numeraire. If the time horizon for protection is T, then the time-zero value of the CVA is

$$
\begin{aligned}
\mathrm{CVA}_{0,T} &= (1 - R_V) \int_0^T \mathbb{E}\left[N_u^{-1} V_u^+ \delta_{\tau - u} du \right] \\
&= (1 - R_V) \int_0^T \mathbb{E}\left[N_u^{-1} V_u^+ \big| \tau = u \right] \mathbb{N}\left(\tau \in (u, u + du) \right),
\end{aligned}
\tag{14.1}
$$

where, δ_x is the Dirac delta at x. We can rewrite this equation as

$$
\begin{aligned}
\mathrm{CVA}_{0,T} &= (1 - R_V) \int_0^T \mathbb{E}\left[D_{0,u}^{-1} N_u^{-1} V_u^+ \big| \tau = u \right] \left\{ D_{0,u} \mathbb{N}\left(\tau \in (u, u + du) \right) \right\} \\
&= \frac{1 - R_V}{1 - R} \int_0^T \mathrm{EPE}_u^{mod} d\mathrm{CDS}_u,
\end{aligned}
\tag{14.2}
$$

where, $D_{0,t}$ is the zero bond price maturing at t, and CDS_t is the value of the protection leg of a credit default swap on the counterparty, maturing at t and with recovery R, where we recall the definitions

$$CDS_t := (1 - R) \int_0^t D_{0,u} \mathbb{N} (\tau \in (u, u + du)), \tag{14.3}$$

$$EPE_u^{mod} := D_{0,u}^{-1} \mathbb{E} \left[N_u^{-1} V_u^+ | \tau = u \right]. \tag{14.4}$$

Consider (14.2): by discretizing the integral it seems natural to try to replicate the credit exposure of the transaction V by a portfolio of positions in CDS contracts of different maturities; (14.2) represents, however, only the net *present* value of buying protection against the counterparty defaulting on the transaction. As the credit spread of the counterparty moves and the price V_t of the transaction evolves, the value of CVA will also change in time. Thus, with a portfolio of CDSs, the value of protection would be equal only to the original expected exposure of the transaction, and *not* to the actual mark-to-market value at default. Moreover, daily CVA changes due to market moves other than moves in the counterparty's credit spreads would not be covered by the hedging portfolio, thereby introducing considerable P&L volatility.

14.2 Contingent Credit Default Swap

In the previous section we have shown what, at trade inception, should be the fair value of a contract compensating for the loss upon default of the counterparty. We have also seen that simply investing this amount in a CDS portfolio, even if continuously adjusted to new CVA values, will not guarantee perfect replication of the credit exposure at time of default.

In order to obtain a portfolio which replicates dynamically the credit exposure of the transaction, consider the evolution of CVA over time. Assuming the process of the portfolio value to be V_t, we can write[3]

$$CVA_{t,T} = (1 - R_V) N_t \int_t^T \mathbb{E} \left[N_u^{-1} V_u^+ \delta_{\tau - u} \mid \mathscr{F}_t \right] du. \tag{14.5}$$

The process $(CVA_{t,T})_{t \geq 0}$ is the price process of a new product, which we refer to as a *contingent credit default swap* (C-CDS) and which at the default time τ of the counterparty has value exactly equal to V_τ^+. The C-CDS is nothing but a derivative on the underlying price process V_t (the value process of the transaction being protected). Our goal is therefore to apply to V_t hedging strategies commonly used for other types of derivatives.

---

[3]Note that the expectation is in the joint law of τ and V, as the two may not be independent.

14.2.1 American Monte Carlo Valuation

In Chap. 5 we introduced the concept of super-product, whose value depends on that of an underlying portfolio of transactions, and we mentioned that an example of super-product is the C-CDS. We also mentioned that valuing a super-product involves computing

(i) the price distribution of the underlying portfolio, V_t,
(ii) the numeraire distribution N_t, and
(iii) the default times τ of the counterparty.

This valuation can be performed by simulation and by using the American Monte Carlo (AMC) technique we described in Chap. 4, for pricing. Consider as an example the C-CDS described in (14.5). Using the price distribution V_t and the default times τ we can build the payoff of the C-CDS on a discretized set of payment dates (T_i).

$$\Pi_{CCDS}(T_i) = V_{T_i}^+ \mathbb{1}_{\tau \in]T_{i-1}, T_i]}. \tag{14.6}$$

To estimate the future price distributions of the C-CDS using the payoff matrix defined above, we need a set of observables, which should at least include

(i) the simulated counterparty CDS spreads, and
(ii) the price distribution V.

It is important to note here that valuing the C-CDS via Monte Carlo allows to automatically take into account the correlation between V and τ. In other words, within the simulation, the distribution of prices is computed conditional on default of the counterparty.

14.2.2 Example

The most straightforward example of a C-CDS is in the case where the underlying portfolio consists of a single interest-rate swap. We will consider this case again at a later stage in this chapter. Let T be the maturity of the swap, K the fixed rate, and $A_{t,T}$ and $s_{t,T}$ respectively the annuity and par rate of the swap seen at time t, and assume that we pay fixed and receive floating.

The payoff of the C-CDS is this case is simply,

$$\Pi_{CCDS}(T_i) = A_{T_i,T} \left(s_{T_i,T} - K \right)^+ \mathbb{1}_{\tau \in]T_{i-1}, T_i]}. \tag{14.7}$$

As an illustration, consider a ten-year USD swap written on a notional of one billion USD, in which we receive the three-month USD Libor rate fixed in advance and pay a fixed coupon equal to today's par rate. Assume the counterparty's CDS curve is flat at 130 basis points. Figure 14.1 displays the profile of the resulting C-CDS. As we can see, the initial point is equal to today's CVA, around 8.4 million

USD.[4] However, because of the underlying interest rate and spread risk, the CVA could reach, at a 97.5% confidence level, up to 22 million USD.

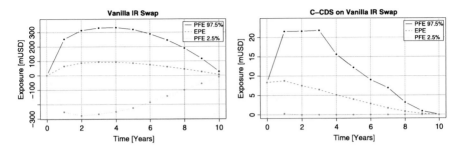

Fig. 14.1 Exposure of a 10 years USD swap, paying quarterly fix par rate and receiving quarterly Libor rate and of the corresponding C-CDS (computation performed August 09)

14.3 Dynamic Hedging of Counterparty Risk

Hedging of counterparty exposure can be performed using C-CDS contracts, which in turn allow hedging strategies common to other types of derivatives. Consider a portfolio Π of hedging instruments $\{H_i\}$, $i = 1, 2 \ldots, n$, with time-t price

$$\Pi_t = \sum_{i=1}^{n} a_i(t) H_i(t), \quad t \geq 0, \tag{14.9}$$

where the coefficients $\{a_i\}$ determine the amounts to be held of each hedging instrument. The self-financing requirement for Π is then given by

$$d\Pi_t = \sum_{i=1}^{n} a_i(t) dH_i(t), \quad t \geq 0. \tag{14.10}$$

Consider for example a transaction of maturity T that depends on only one risk driver, say, the EUR rate of interest. Suppose that suitable hedging instruments are

---

[4] A useful CVA approximation for an at-the-money swap, (see Appendix A), is

$$CVA_T \simeq \frac{2T}{3} EPE_{max} s_T, \tag{14.8}$$

where $T = 10$ years and s_T is the CDS spread on the counterparty. Looking at the profile of the swap, we can see that the maximum EPE is roughly 100 million USD, hence the approximative computation gives us roughly the right result, since $100 \times 2/3 \times 0.013 \times 10 \simeq 8.6$.

a bank account $\{B_t\}$, a forward (i.e. swap) of price process $\{F_t\}$, and an option (i.e. swaption) of price $\{O_t\}$, all with the EUR rate of interest as underlying:

$$\Pi_t = a(t)F_t + b(t)O_t + c(t)B_t. \tag{14.11}$$

The holding coefficients a and b would typically be chosen to make the portfolio Π delta and gamma flat, that is,

$$\begin{cases} a(t) = \Delta - \frac{\Gamma}{\Gamma_O}\Delta_O \\ b(t) = \frac{\Gamma}{\Gamma_O}, \end{cases} \tag{14.12}$$

where Δ and Γ (resp. Δ_O and Γ_O) represent the first and second order derivatives of the transaction price (resp. the hedging option) with respect to the underlying risk driver.

In the more general case where the transaction being hedged depends on $k > 1$ risk drivers (X_j), $j = 1, 2, \ldots, k$, and assuming that a correct hedge can be achieved using a forward and an option for each risk driver, we have $n = 2k$ and

$$\begin{cases} \Pi_t & = \sum_{i=1}^{2k} a_i(t)H_i(t) \\ H_{2(j-1)+1}(t) & = F_j(t) \\ H_{2(j-1)}(t) & = O_j(t) \\ a_{2(j-1)+1}(t) & = \Delta^{(j)} - \frac{\Gamma^{(j)}}{\Gamma_O^{(j)}\Delta_O^{(j)}} \\ a_{2(j-1)}(t) & = \frac{\Gamma^{(j)}}{\Gamma_O^{(j)}}, \end{cases} \tag{14.13}$$

where the subscript now indicates derivatives with respect to the j'th risk driver.

The hedging strategy described by (14.13) can be carried through when the transaction to be hedged is a C-CDS. Because one of the main underlying risk factors is the credit quality of the counterparty, the hedging portfolio Π will have holdings in CDS contracts on that counterparty. In addition it will also be necessary to trade instruments that depend on the market risk factors driving the exposure in the underlying portfolio. Finally, it will be necessary to consider correlation risk between different risk drivers. Because of the lack of instruments that can cover second order derivatives, the so-called cross-gammas, this risk is in general difficult to hedge.

Putting in place the hedge (14.13) for the C-CDS requires estimating sensitivities (first and second derivatives) of the C-CDS contract with respect to several risk factors. How to do this is what we discuss next.

14.4 Optimal Static Hedging

In the last section we discussed dynamic hedging of a derivative from a theoretical standpoint. In practice, rebalancing the replicating portfolio may be expensive and,

as we mentioned already, the hedging instruments could be illiquid, or, in a financial crisis, liquidity could dry up. As an alternative we therefore briefly investigate,
the possibility of determining an *optimal static hedge*. As for the C-CDS, this will
involve the usage of price distributions of other financial products, creating what we
have called super-products.

Assume to have n hedging instruments available. Define $H_i(t)$ the present value
of the i'th hedging instrument at time t, and α_i the amount of hedging instrument
held. Because the future values, CVA(t), of the C-CDS we are trying to replicate
are known along with those of each hedging instrument, we know the future values
of the P&L resulting from our hedging strategy $\alpha \equiv (\alpha_i)$. One strategy could be for
example, to try to maximise the profits resulting from our hedging, while minimising
its overall variance. To this end, define the numeraire-scaled gain process of our
strategy by

$$
G_\alpha(t) = N_t^{-1} \left[\sum_{i=1}^n \alpha_i \left(H_i(t) - H_i(0) N_t \right) - \left(\text{CVA}(t) - \text{CVA}(0) N_t \right) \right]^+ , \quad (14.14)
$$

and $M_\alpha(t, k)$ as the k'th moment of $G_\alpha(t)$,

$$
M_\alpha(t, k) = \mathbb{E}\left(G_\alpha(t)^k \right), \quad k = 1, 2, \ldots . \quad (14.15)
$$

One possibility is then for example to try maximising the Sharpe ratio of the
P&L, defined as

$$
\max_\alpha \left(\frac{M(t, 1)}{\sqrt{M(t, 2) - M(t, 1)^2}} \right). \quad (14.16)
$$

Clearly determining the optimal strategy will have constraints in the available
instruments and could be computationally intensive as it requires the usage of full
portfolio price distributions over time. This approach, however, is appealing as (i)
we can at least in theory define several objective functions, and not just ensure that
we are 'delta-flat', (ii) as the hedge is built to be static, it minimises rebalancing requirements, and (iii) it can be used in conjunction with dynamic hedging strategies.

14.5 CVA Sensitivities

In general, estimation of price sensitivities to underlying risk factors can be problematic from both the algorithmic and computational point of view, as it is performed by taking numerical derivatives of pricing functions. As we have seen in
Chap. 4, when pricing using AMC techniques, the required desired perturbation of
the price distribution can be achieved by estimating price distributions at two time
points, say t and $t + \varepsilon$ with ε small. Because AMC estimates the price distribution
as a function of chosen underlying stochastic variables, it is then straightforward
to write down derivatives of the price distribution with respect to these variables.

The only restriction is of course that sensitivities with respect to a non-stochastic quantity cannot be calculated in this way.[5]

This approach to computing sensitivities is appealing for at least two reasons: (i) computational speed (only one extra regression step is required), and (ii) flexibility (sensitivities to any function of the risk drivers can be computed). Nevertheless, one has to exercise care in using this approach, which gives, by construction, the sensitivity with respect to a chosen observable (function of the underlying risk drivers). If sensitivity to an individual risk factor, keeping all other factors constant, is required, one needs to go through the *de-correlation* process described in Sect. 4.5.3.

There are several components of CVA sensitivities. We describe four of them here.

(i) *Credit spread.* This is in general the main source of CVA change. This risk can be hedged reasonably well provided that there is enough liquidity on the counterparty CDS. Typically, the credit spread risk is decomposed into chosen time buckets in order to match the limited range of CDS instrument maturities that are available in the market.

   The estimation of credit spread sensitivities themselves could be done reasonably efficiently by employing finite-difference approximations on the expression (14.2).

   In cases where the liquidity for a given counterparty is poor or non-existent, protection is bought on indices with the intention of covering generic or sector risk. Idiosyncratic risk will remain un-hedged.

(ii) *Underlying Market Risk.* The underlying transaction (of price process V) that gives rise to the credit exposure portfolio may be composed of all kinds of products, across all assets classes, of all levels of complexity. Thus, any move in a market parameter (e.g. a move in the 10Y USD swap rate, the EUR/USD rate, the JPY/USD volatility skew or the iTraxx correlation...) will affect the Modified EPE and therefore the value of the C-CDS. Most of this risk should, if possible, be hedged by transacting exchange-traded futures and options, therefore avoiding adding risk to another counterparty.

(iii) *Counterparty—Portfolio Correlation Risk.* This is the interplay between market and credit risk, the dependence between the event of counterparty default and the value of the transaction that gives rise to the credit exposure. We have discussed, in Chap. 13, two technical solutions that take into account this dependence by modifying the price distribution after this has been estimated. The real question is how to quantify the level of dependence involved. In the relatively simple cases (e.g. if the underlying portfolio consists mostly of credit derivatives on names which are strongly correlated with the counterparty) the problem is similar to having to deal with correlation between assets in the same asset class. In most cases, however, there can be several ways to represent market/credit correlation, all of which are realistic, but all of which lead to different

---

[5]Note that to compute sensitivities by regression we only need to have a price distribution generated with the relevant stochastic variables. This distribution could be obtained with techniques different from AMC.

C-CDS prices. This makes correlation risk very difficult to hedge in general, and one often deals with this by studying a range of possibilities rather than choosing one particular value of market/credit dependence.

The difficulties above are mitigated somewhat by the fact that market/credit correlation is rarely the main risk driver in practice. For example, in the case of a portfolio consisting of interest rate or foreign exchange products, the dependence between counterparty credit quality and the underlying interest/exchange rates may safely be ignored unless the counterparty happens to be a sovereign entity.

(iv) *Break Clauses.* Break clauses are often embedded in transactions to limit the length of time for which counterparties are exposed to changes in the transaction value. The effect of a break clause is obviously to reduce the value of a C-CDS contract (since there are no cashflows to be protected beyond the break date). On a portfolio level, different break clauses pertaining to the constituent trades may have different break dates; the net effect of such break clauses will then be an amortisation of the portfolio notional.

Very often, CVA is computed initially by taking break clauses into account, but for several reasons (not least, to maintain healthy relationships with clients) it may happen that some break clauses are not exercised when triggered. At such break dates where break clauses are not invoked, extra credit protection will need to be bought and the CVA recalculated.

14.6 Collateral Agreements

Section 12.4 discussed the mitigation of credit exposure by using collateral agreements. We discuss here the effect that such agreements have on the calculation of CVA.

It is immediate that a collateral agreement serves to reduce the exposure faced on a transaction from V_t to $(V_t - C_t)$, where C_t is the value of collateral. For instance, if margin calls are allowed daily, and the threshold is zero, then the daily exposure on the trade is made to vanish. Implicit in this remark is the assumption, of course, that the holder of the collateral is able, at the time of counterparty default, to instantly liquidate the collateral held. If the collateral is in the form of non-cash assets, redeeming it in the form of cash will take time, during which the value of the underlying transaction is bound to change. Any changes will be borne by the collateral holder in the form of close-out risk.

In other words, while the CVA of a transaction is reduced by holding collateral, it cannot be made to vanish entirely because losses can be incurred while liquidating the collateral at the point of default. This becomes all the more important when the collateral consists of non-liquid assets, for then the close-out risk is borne over a longer period of time. We will illustrate below, through a specific example, how the presence of close-out risk makes difficult even the calculation and hedging of CVA for linear products.

14.7 Right-Way/Wrong-Way Risk

The reader will recall, from Sect. 13.4, the effect on credit exposure of dependence between the underlying portfolio and the event of default of the counterparty. Now, because the C-CDS derives its value from the probability of the counterparty defaulting, the effect of dependence between counterparty default probability and the underlying portfolio distribution will be felt most acutely when the underlying portfolio itself contains credit-related instruments. For instance, upon default of the counterparty in such a transaction, credit spreads of all correlated entities present in the portfolio are bound to change significantly, causing an immediate change in the value of the underlying transaction.

In our framework, dependence between credit entities is specified as in (2.72), and such dependence will automatically be reflected in any C-CDS valuation where the underlying portfolio contains credit instruments. Hedging the C-CDS in such a situation is a difficult task—in principle one would expect to have to include correlation-sensitive instruments such as CDO tranches that involve both the counterparty and the underlying entities.

In principle we could extend our modelling framework to take into account dependence between credit and other asset classes, thus incorporating right-way/wrong-way risk in the CVA computation. Note that this would not replace the approach described in Chap. 13. In this case the focus was on computing the full portfolio distribution, conditional on default of the counterparty, and not simply the value of the C-CDS. The two approaches could be reconciled by extracting paths in the simulation that correspond to a default of the counterparty. Given that defaults are a rare event, this would require a much higher number of simulations to provide meaningful results.

14.8 Examples

We now discuss some examples that illustrate the theoretical aspects we have examined above.

14.8.1 C-CDS on a Vanilla Interest-Rate Swap

Consider a corporate which, in order to hedge its exposure to USD interest rates, enters into a large ten-year interest rate swap with an Investment Bank. In a financial downturn we can imagine the corporate being concerned about the solvency of its counterparty on the trade, and we could think that it considers buying from a better-rated institution a C-CDS on this swap.

In this case, assuming no correlation between the USD rates and defaults, it is relatively straight-forward to replicate a C-CDS on this swap by using swaptions and CDSs. The value of the C-CDS at time t can be written as,

$$\text{CCDS}_t = \int_t^T \text{Swaption}(t, u, T) \mathbb{N}(\tau \in (u, u + du) \mid \tau > t), \qquad (14.17)$$

where T denotes the maturity of the underlying swap, and $\text{Swaption}(t, u, T)$ denotes the value at t of a swaption expiring at u with underlying swap maturity T, struck at the underlying swap's fixed rate. By discretizing this expression, we can write

$$\text{CCDS}_t \simeq \sum_{t_i > t}^n \text{Swaption}(t, t_i, T) P(t, t_{i-1}, t_i), \qquad (14.18)$$

where $P(t, t_{i-1}, t_i) = \mathbb{N}(t_{i-1} \leq \tau \leq t_i \mid \tau > t)$. The value of the C-CDS at time t can therefore be replicated either as

(i) the value of an amortising CDS with final maturity T and notional equal to the modified EPE profile seen at time t as per (14.2), or as

(ii) the value of a strip of swaptions with underlying maturity T and with individual notional equal to the incremental default probabilities.

This points to the fact that any dynamic hedging of the C-CDS contract will involve purchasing both interest-rate swaptions and CDS contracts.

14.8.2 Impact of Discretization Schedule

The choice of the discretization schedule can substantially alter the CVA value, in particular when cashflows are paid by the underlying instrument. Consider for example the swap of the previous section and assume that the discretization schedule coincides with the coupon payment dates. The figure below depicts the EPE profile of the swap taking into account the yearly coupons. The upper curve shows an hypothetic profile computed assuming that coupons are paid just after valuation points. This schedule would over-estimate the cost of hedging. On the other side a discretization profile with dates just after coupon payments would under-estimate CVA, as it is shown in the lower curve.

In this case to take into account coupons and correct the CVA valuation, we could include in the computation a sum of caplets,

$$\text{CCDS}_t \simeq \sum_{T_i > t}^n \text{Swaption}(t, T_i, T) P(t, T_{i-1}, T_i)$$

$$+ \sum_{T_i > t}^n \text{Caplet}(t, T_i) P(t, T_{i-1}, T_i), \qquad (14.19)$$

where (T_i) is the set of coupon dates.

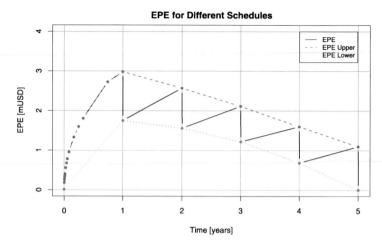

Fig. 14.2 Exposure profile of a swap paying yearly coupon on a notional of 100 mUSD. The upper (resp. lower) curve shows an hypothetical profile with coupon payments just after (resp. before) discretization schedule

14.8.3 Collateralised Equity Swap

The example we give now is intended to highlight how even simple products can give rise to C-CDS contracts that present a non-trivial hedging problem.

Consider a forward contract on a non-dividend-bearing stock that is non-liquid, and suppose that the counterparty agrees to collateralise this transaction with daily margin calls and zero threshold.

While the collateral agreement serves to maintain the value of collateral equal to the credit exposure of the forward at all times, the lack of liquidity means that upon default, the collateral holder has to contend with close-out risk. If we write S_t for the value process of the stock, and assuming a close-out period of length $\delta > 0$, then the time-zero value of the close-out risk per share of stock faced at default is given by

$$\mathbb{E}\left[N_t^{-1} \left(S_t - S_{t-\delta} \right)^+ \right], \tag{14.20}$$

therefore hedging the close-out risk on a stock forward involves hedging the cliquet-type payoff appearing in the expectation in the equation written above.

14.9 Case Study

To highlight the importance of dynamically hedging CVA, consider the following stylised example. Suppose that in June 2008 we had entered into a 20-year EUR/USD forward contract of 500 million EUR notional, in which we receive USD and pay EUR. To value the trade at par, the contract would have been struck at around 1.65 USD per EUR.

Assume now that the counterparty we are facing had at inception a CDS curve trading at 300 bps flat. The initial CVA for this transaction would have been in the order of 10 mUSD. Leaping one year forward to June 2009, because of the fall in the EUR/USD exchange rate, the forward contract is now worth about 47 mUSD in our favour.

Figure 14.3 shows the difference in exposure profiles generated on both dates. As we can see, the EPE profile computed in June 2009 is substantially higher than that computed a year before.

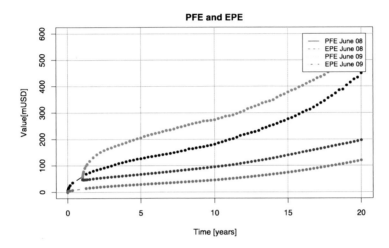

Fig. 14.3 PFE and EPE profiles computed June 08 and June 09. The two set of profiles are superimposed and the time axis is referred to the 2008 computation. The 2009 EPE and PFE profile start at the one-year point. Note that their initial value is within the PFE profile computed one year earlier, but due to the new market condition, the new profiles are outside the bounds computed in 2008

It is also interesting to note that the present value of the trade in June 2009 lies within the PFE confidence level computed in 2008.[6] Assuming that the counterparty spread stayed at 300 bps, the resulting CVA would have gone up to roughly 24 mUSD. Had we only hedged the risk relating to the counterparty's CDS curve, we would therefore face a loss of about 14 mUSD. Figure 14.4 shows the difference in CCDS profiles generated on both dates.

Assume now that at inception of the trade in 2008, we had decided to also hedge the EUR/USD risk. To do so, we would have needed to choose an instrument which is liquidly traded and that does not add additional counterparty risk, such as one-year EUR/USD futures. In June 2008, the EUR/USD delta for the CVA of our fictional

---

[6]Note that the EUR/USD exchange rate saw its greatest historical absolute fall between 2008 and 2009.

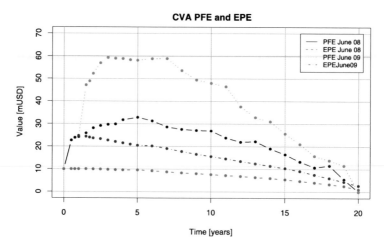

Fig. 14.4 CVA EPE and PFE profiles computed in June 08 and 09. The initial points of the two sets of profile corresponds to the CVA computed at inception and one year later

transaction was of the order of −37 mUSD, meaning that for every 0.01 move in the EUR/USD exchange rate, the CVA would increase by 370 kUSD. Hence, being long 37 mEUR notional worth of one-year futures should in theory eliminate the currency risk.

The table below summarises the result of our hedging strategy from inception until June 2009. As we can see, the hedging strategy would have resulted in a profit of roughly 1 mUSD, as opposed to an un-hedged loss of 14 mUSD.[7]

Table 14.1 Summary of hedging strategy including CDS and FX hedges

	June 2008	June 2009
CVA	−9,983,526	−24,061,324
Hedge	0	14,870,065
Cash	9,983,526	10,185,207
Net	0	993,948

Of course, it is unrealistic to assume that the counterparty spread would have remained unchanged during a one year period. Assume now that in fact the CDS spread would have increased from 300 bps to 400 bps. In this case, the CVA computed in 2009 would no longer be 24 mUSD, but 27.7 mUSD. The credit delta[8] computed in 2008 was of the order of 20 kUSD per basis point. While the maturity of the underlying portfolio is 20 years, assume that the only liquid maturity for the counterparty CDS is five years, for which the credit delta would be of the order of

---

[7]Note that we have assumed that the deposit rate at which the cash grows is 2% per annum.

[8]In other words, the counterparty spread delta of the CVA.

400 USD per basis point on a one million USD notional, meaning that in order to hedge the counterparty spread risk we should enter into a 5 year CDS of roughly 50 mUSD notional. In June 2009, this position would have yielded a profit in the region of 1.6 mUSD, partially offsetting the loss.

We can see now how, even in this stylised example, a simple market hedge can considerably improve P&L resulting from changes in CVA. An un-hedged position would have resulted in a net loss of almost 18m USD, while the full hedge we described would have reduced this loss to about 1 mUSD. It is worth noting that a static hedge involving solely the counterparty CDS would have resulted in a loss of roughly 16 mUSD.

From this and the previous examples we can see that replicating a C-CDS involves hedging a hybrid product, which has market, (e.g. FX or interest rate), and credit components. Ignoring for example the FX risk would clearly undermine any hedging strategy. It is interesting to note that in a classical set-up, where the CVA is computed statically using simply the EPE profile and the spread of the counterparties, the market risk components of the hedge are difficult to compute, as they require the EPE sensitivities to market risk factors. They involve, however, the usage of instruments, which are in general traded on exchanges. The credit component can be computed more easily, but on the other side it involves CDS products, which are still mainly traded as OTC transactions and are not available for all names and all maturities.[9] This involve the usage of credit indices and the finding of curves which can be used as proxies for illiquid names.

In addition to the risks we have already mentioned, we need to consider the so called *vega-risk* deriving from movements of implied volatility. This can be substantial especially when the portfolio is dominated by FX positions. To appropriately hedge this risk we need to include in the C-CDS computation volatility as a stochastic driver. This can be performed by implementing a stochastic volatility model, as described in Chap. 3.

---

[9]To reduce the counterparty risk in general CDS hedges are traded with fully collateralised counterparties. There are extensive discussions to standardise CDSs and trade them on exchanges.

Concluding Remarks

Our goal in this book was to model counterparty credit exposure for all types of transactions. We saw that by appropriately choosing the fundamental quantities to model we can approach the problem in a modular way, dividing features and conquering products.

Price distributions are obtained using American Monte Carlo (AMC) techniques, allowing a valuation framework where modularity and flexibility are key. With the introduction of a booking language, PAL, we added a further layer of de-coupling and abstraction, enabling a system architecture that could address most of the problems faced by a counterparty exposure system dealing with large diverse portfolios.

The natural next step was to investigate how to manage counterparty exposure, both in static and dynamic ways. This led to the introduction of the so-called contingent credit default swap product, C-CDS, which replicates the cost of protection.

We now summarise the steps needed to compute and hedge credit exposure.

(i) *Translation.* First of all, all trades within the portfolio should be understood by the valuation engine. This means that each trade needs to be translated into the common trade representation language.

(ii) *Portfolio Valuation.* Once the first stage is completed, it is possible to model the underlying risk drivers, which have been recognised via the common trade representation language, and value each trade, along with its future price distributions. All trades are then aggregated together, including possible netting rules or break clauses, to finally arrive at the future distributions of the portfolio. If a collateral agreement exists, its logic should then be applied to the portfolio distributions.

(iii) *C-CDS Valuation.* The credit valuation adjustment, CVA, can be valued using the modified EPE profile of the portfolio and the counterparty credit spread curve. Using C-CDSs, however, we can compute not only the value of CVA, but also the CVA future price distribution.

(iv) *Sensitivities Computation and Replication.* As a final step, sensitivities can be computed from the C-CDS distribution, using either a regression-based approach, or a full revaluation (known in the industry as 'bumping' method), starting the process again from step (ii).

G. Cesari et al., *Modelling, Pricing, and Hedging Counterparty Credit Exposure,*
Springer Finance, DOI 10.1007/978-3-642-04454-0,
© Springer-Verlag Berlin Heidelberg 2009

(v) *Post-processing.* For purposes of risk control (e.g. to compute regulatory capital or compare PFE with limits), a post-processing of the price distribution may be needed. Examples of this include stress-testing and accounting for right-way/wrong-way risk.

The techniques we described can also be applied to other problems that large financial companies need to address. Examples are (i) computing the value of the so called *own credit* of a company, (ii) valuing *debt valuation adjustments* (DVA) of portfolios of transactions, (iii) addressing the problem of valuing the *cost of funding* and *cost of collateral*, (iv) computing potential values of transactions in different scenarios, (v) determining the value of *risk weighted assets* and of *regulatory capital*, or (vi) investigating various hedging strategies. All these problems deserve a thorough analysis which could be the subject of further research. It is interesting to note here that any solution to these questions will require, as fundamental feature, the capability of computing future distributions of prices. This is the feature at the heart of our work.

A final remark to conclude. What we described in this work is only a brief overview of the problem we try to solve. As we highlighted throughout this book, in many occasions we accepted compromises in our implementation and highlighted shortcuts. Many points can be improved, further explored and changed. We think, however, that at a general level, the framework and the ideas we provide are a viable solution to the modelling, pricing and hedging of counterparty credit exposure for large portfolios of different products.

Appendix A
Approximations

We summarise here some useful approximations of counterparty exposure computation, often used by practitioners. While they cannot provide satisfactory results in general, they may serve as a sanity check for more complex computations, and to help intuition. In some cases in the computation of Expected Positive Exposure (EPE) for some types of products, they are based on pricing information and give exact valuation. Some of the formulae we present are general and others can be used only for specific products. We consider here what we found useful in our day-to-day work.

A.1 Maximum Likely Exposure

In general, the Potential Future Exposure profile (PFE) of a given product is a function of time. We call its maximum value Maximum Likely Exposure (MLE). In the following sections we provide some MLE estimate for simple products.

A.1.1 MLE of Equity and FX Products

MLE values can be easily approximated in the case of options or forwards on assets that can be modelled as Geometric Brownian Motions assuming constant volatility and interest rate. Under these assumptions in fact the exposure profile reaches its maximum at maturity of the trade, where its value coincides with its intrinsic value. Thus, to compute the MLE, what is necessary is to estimate the potential value of the asset at maturity.

Consider for example an option on a stock S with Black-Scholes volatility σ, interest rate r, and strike K. The maximum value of the exposure at maturity T (within a 97.5% confidence interval) is given by

$$\text{MLE} = Se^{(r-\frac{1}{2}\sigma^2)T+1.96\sigma\sqrt{T}} - K. \tag{A.1}$$

G. Cesari et al., *Modelling, Pricing, and Hedging Counterparty Credit Exposure,* Springer Finance, DOI 10.1007/978-3-642-04454-0, © Springer-Verlag Berlin Heidelberg 2009

If we assume zero interest rate, stock returns normally distributed, and at the money products ($S = K$), we can simplify this formula as follows,

$$\text{MLE} = 1.96 S \sigma \sqrt{T}. \tag{A.2}$$

The main problem in these valuations is the choice of volatility. If the volatility is assumed to be constant, it is necessary to estimate the value that will best fit the terminal asset distribution. If the choice is to use implied volatilities, the at-the-money volatility is often the most suitable one to use. In practice if implied volatilities are not available historical volatilities are used.

A.1.2 MLE of Swaps

Throughout this book we have seen several PFE profiles of interest-rate swaps. In general, when the product is vanilla, they show a typical bell shape, which starts from zero, increases over time and then decreases to reach zero again at maturity. This shape is driven by two factors, the declining duration (time to maturity) and the increasing variance of the swap. Assume that, at any time t, the duration is proportional to the remaining life of the swap via a constant $A_0 < 1$, and that the interest-rate volatility increases with the square root of time, $\sigma_N \sqrt{t}$.[1] We can write the volatility of the swap as

$$\text{Vol}_{Swap} = A_0(T - t)\sigma_N \sqrt{t}. \tag{A.3}$$

The peak exposure, i.e. the MLE, is reached at about one third of the life of the trade. We can see this by simply taking the first derivative of the volatility with respect to time, and imposing its value to be zero.

$$\frac{\partial \text{Vol}_{Swap}}{\partial t} = 0 \Longleftrightarrow -A_0 + A_0(T - t)\frac{1}{2t} = 0 \Longleftrightarrow t = \frac{T}{3}. \tag{A.4}$$

Using this result and assuming that the price distribution of an at-the-money swap is normally distributed, we can estimate the price distribution of a swap at time $T/3$,

$$\text{SwapDistribution}_{t=T/3} \approx A_0 \frac{2}{3} T \sigma_N \sqrt{\frac{T}{3}} Z, \tag{A.5}$$

where $Z \sim N(0, 1)$. If we want to value the MLE, i.e the peak PFE exposure at 97.5% confidence interval, we need to substitute Z with 1.96. The present value of EPE can be computed by taking the expectation of the positive part of this distribution. Doing this we obtain

$$\text{EPE}_t^{PV} \approx \frac{1}{\sqrt{2\pi}} A_0(T - t)\sigma_N \sqrt{t} \approx 0.4 A_0(T - t)\sigma_N \sqrt{t}. \tag{A.6}$$

---

[1] σ_N is the volatility of a normal distribution. It is related to the log-normal (Black-Scholes) volatility σ of the swap rate via the level of interest rate, $\sigma_N \approx r\sigma$.

A.2 Expected Positive Exposure

The Expected Positive Exposure (EPE) computation is strongly related to pricing. In general, under pricing measure assumptions, the EPE of a transaction at time t is the price of an option to enter in the transaction at time t. This is a very useful result, as it allows to approximate EPE computations using price information.

A.2.1 EPE and CVA of Equity Options

As a first example consider an option on a stock or an FX currency. Under simplified assumptions, EPE can be written as

$$EPE_t = \mathbb{E}[V_t^+], \qquad (A.7)$$

where V_t is the price distribution at time t. In the case where V_t is always non-negative, as for example for options, this equation becomes

$$EPE_t = \mathbb{E}[V_t^+] = \mathbb{E}[V_t] = \mathbb{E}[\mathbb{E}[e^{-r(T-t)}(S_T - K)^+|\mathscr{F}_t]] = V_0 e^{rt}, \qquad (A.8)$$

where we have assumed constant interest rates and volatility. Thus,

$$EPE_t = V_0 e^{rt}. \qquad (A.9)$$

In other words the EPE of an option at time t is the option premium increased at the risk-free rate.

The CVA can be computed as the discounted EPE multiplied by the spread (assumed to be constant) multiplied by time to maturity,

$$CVA \approx V_0 s_0 T. \qquad (A.10)$$

This formula holds for any product whose price distribution is non-negative and which does not pay intermediate cashflows. For example it can be used to compute CVA of a cash-settled swaption, while it cannot be applied in the case of a physically-settled swaption.

A.2.2 Relation between MLE, EPE

If we assume zero interest rate we can approximate the price V_0 of an at-the-money ($S = K$) option as,

$$V_0 = S\Phi(d_1) - K\Phi(d_2), \qquad (A.11)$$

where Φ is the cumulative normal distribution, and

$$d_{1/2} = \frac{\ln(S/K)}{\sigma\sqrt{T}} \pm \frac{\sigma\sqrt{T}}{2} = \pm\frac{\sigma\sqrt{T}}{2}. \tag{A.12}$$

Using the following approximation

$$\Phi(x) = \frac{1}{2} + \frac{1}{\sqrt{2\pi}}x + O(x^3), \tag{A.13}$$

and assuming $\sigma\sqrt{T} \ll 1$ the above equation becomes,

$$V_0 = S\Phi\left(\frac{\sigma\sqrt{T}}{2}\right) - K\Phi\left(-\frac{\sigma\sqrt{T}}{2}\right) \approx \frac{1}{\sqrt{2\pi}}S\sigma\sqrt{T} \approx 0.4S\sigma\sqrt{T}. \tag{A.14}$$

We have seen that the EPE of an option can be computed as the option premium growing at risk free rate. Thus

$$\mathrm{EPE} \approx 0.4\sigma\,S\sqrt{T}. \tag{A.15}$$

We can now compute a relation between EPE and the 97.5% MLE. Recall that,

$$\mathrm{MLE} \approx 1.96\sigma\,S\sqrt{T}. \tag{A.16}$$

Thus, we obtain,

$$\frac{\mathrm{EPE}}{\mathrm{MLE}} \approx 0.2. \tag{A.17}$$

In other words, if the distribution of the portfolio is normal and centered around zero, then the 97.5% MLE is roughly five time larger than the EPE.

A.3 CVA of Swaps

The EPE value at time t of a swaps portfolio is often computed by practitioners as the value of a swaption, i.e. the value of an option to enter into a (portfolio of) swaps. This valuation is correct, however, only if the modified value of the EPE, as defined in Chaps. 12 and 14, is used. Often this valuation methodology is called *swaption approach*.

We can evaluate approximation of the CVA of a swap as follows.

$$\mathrm{CVA}^{swap} \approx \int_0^T \mathrm{EPE}_u^{PV} s_0 du \approx s_0 \int_0^T 0.4A_0(T-u)\sigma_N\sqrt{u}du$$

$$= s_0 0.4A_0\sigma_N T^{5/2}\frac{4}{15}, \tag{A.18}$$

where s_0 is the CDS spread and EPE^{PV} is the present value of the EPE. Recalling (A.5) we can approximate the peak value of the discounted EPE profile as

$$\text{EPE}_{max}^{PV} \approx 0.4 A_0 \sigma_N T^{3/2} \frac{2}{3\sqrt{3}}, \qquad (\text{A.19})$$

and thus,

$$\text{CVA} \approx s_0 T \frac{6\sqrt{3}}{15} \text{EPE}_{max}^{PV}. \qquad (\text{A.20})$$

Noting that the maximum value of the EPE profile of an at-the-money swap occurs at $t = T/3$ and using the 'swaption approach' we defined earlier, we get,

$$\text{CVA} \approx s_0 T \frac{2}{3} \text{Swaption}\left(\frac{T}{3}, T\right), \qquad (\text{A.21})$$

where we have approximated $6\sqrt{3}/15$ with $2/3$, and Swaption$(\frac{T}{3}, T)$ is the value of an option to enter at time $T/3$ into a swap of maturity T.

Appendix B
Results from Stochastic Calculus and Finance

This book is concerned with the pricing and hedging of risk borne by financial institutions when entering into transactions with other counterparties. Such risk arises
from the random nature of the prices of products transacted as well as the possibility
that the counterparty defaults, but its pricing and replication uses the same concepts
as for other kinds of financial derivatives.

This appendix collects a few technical results that we will need throughout. We
start by giving definitions for the basic stochastic processes we use, and then recall
the concept of change of measure. We give also a brief overview of the fundamental
theorem of asset pricing, which allows us to characterise the hedging portfolio for a
traded derivative from martingale representation.

Derivation and analysis of these results can be found in standard finance books,
such as Baxter & Rennie [10], Hunt & Kennedy [64], Karatzas & Shreve [68],
Rogers & Williams [93, 94], Shreve [98], and Williams [106].

B.1 Brownian Motion and Martingales

All our processes are defined relative to a filtered probability space
$(\Omega, \mathscr{F}, (\mathscr{F}_t)_{t \geq 0}, \mathbb{P})$, where $(\mathscr{F}_t)_{t \geq 0}$ is a filtration in $\mathscr{F}$. The basic process we work
with is Brownian Motion.

Definition 1 A process $W \equiv (W_t)_{t \geq 0}$ on $(\Omega, \mathscr{F}, \mathbb{P})$ is called Brownian Motion if

(i) $W_0(\omega) = 0$, for all paths $\omega \in \Omega$;
(ii) for each $\omega \in \Omega$, $W_t(\omega)$ is a continuous function of t;
(iii) for each $t, h \geq 0$, $W_{t+h} - W_t$ is independent of W_t, and has a Gaussian distribution with mean 0 and variance h.

Brownian Motion is an example of a martingale, the most important class of
processes.

Definition 2 A process M is called a martingale with respect to $(\mathscr{F}_t)_{t \geq 0}$ if

G. Cesari et al., *Modelling, Pricing, and Hedging Counterparty Credit Exposure*,
Springer Finance, DOI 10.1007/978-3-642-04454-0,
© Springer-Verlag Berlin Heidelberg 2009

(i) M is adapted, that is M_t is $\mathscr{F}_t$-measurable;
(ii) $\mathbb{E}[|M_t|] < \infty$;
(iii) if $s \leq t$, then $\mathbb{E}[M_t \mid \mathscr{F}_s] = M_s$.

M is a *supermartingale* (resp. *submartingale*) if we replace equality in (iii) above by $\leq$ (resp. $\geq$). For proving general results, the class of martingales is not the right notion to work with, and one needs to consider *local martingales*. While all martingales are also local martingales, the converse is true only if certain conditions hold. The distinction will not be important for our purposes in this book.

At the heart of most of what we do is the idea of looking at various processes in a measure different from that of the given probability triple $(\Omega, \mathscr{F}, \mathbb{P})$. Indeed, if Z is a non-negative random variable (that is, $\mathscr{F}$-measurable) then

$$\tilde{\mathbb{P}}(F) := \mathbb{E}[Z \mathbb{1}_F]/\mathbb{E}[Z], \quad F \in \mathscr{F} \tag{B.1}$$

defines a new probability measure $\tilde{\mathbb{P}}$ on $\mathscr{F}$ for which

$$\mathbb{P}[F] = 0 \implies \tilde{\mathbb{P}}[F] = 0. \tag{B.2}$$

The last implication allows us to make the following definition:

Definition 3 A probability measure $\tilde{\mathbb{P}}$ on $(\Omega, \mathscr{F})$ is said to be *absolutely continuous* with respect to $\mathbb{P}$, denoted $\tilde{\mathbb{P}} \ll \mathbb{P}$, if for all $F \in \mathscr{F}$, (B.2) is true. If both $\tilde{\mathbb{P}} \ll \mathbb{P}$ and $\mathbb{P} \ll \tilde{\mathbb{P}}$ are true, then $\mathbb{P}$ and $\tilde{\mathbb{P}}$ are said to be equivalent. In this case, $\mathbb{P}$ and $\tilde{\mathbb{P}}$ have the same sets of measure zero.

The converse to (B.1) is given by the Radon-Nikodym theorem.

Theorem 1 Let $\tilde{\mathbb{P}} \ll \mathbb{P}$ be a probability measure that is absolutely continuous with respect to $\mathbb{P}$. Then $\tilde{\mathbb{P}}$ can be characterised as in (B.1) for some non-negative random variable Z, which is then called the Radon-Nikodym derivative of $\tilde{\mathbb{P}}$ with respect to $\mathbb{P}$, and we write

$$Z \equiv \frac{d\tilde{\mathbb{P}}}{d\mathbb{P}}. \tag{B.3}$$

The context in which we will most often see measure-change at work is when changing the drift of a Brownian Motion process. Given a $\mathbb{P}$-Brownian Motion W, if the process $\gamma \equiv (\gamma_t)_{t \geq 0}$ is such that

$$\zeta_t := \exp\left\{ \int_0^t \gamma_s dW_s - \frac{1}{2} \int_0^t \gamma_s^2 ds \right\} \tag{B.4}$$

is a martingale, then there exists a unique probability measure $\tilde{\mathbb{P}}$ such that

$$W_t - \int_0^t \gamma_s ds \tag{B.5}$$

is a $\tilde{\mathbb{P}}$-Brownian Motion. Moreover, the Radon-Nikodym derivative of $\tilde{\mathbb{P}}$ relative to $\mathbb{P}$ is given on every $\mathscr{F}_t$ by

$$\left.\frac{d\tilde{\mathbb{P}}}{d\mathbb{P}}\right|_{\mathscr{F}_t} = \zeta_t. \tag{B.6}$$

The above says that under $\tilde{\mathbb{P}}$, W has a drift of γ. Equivalently,

$$\tilde{W} \text{ is a } \tilde{\mathbb{P}}\text{-martingale} \iff \zeta \tilde{W} \text{ is a } \mathbb{P}\text{-martingale}. \tag{B.7}$$

The change-of-measure technique is an indispensable device for simplifying calculations by removing from a process an unwanted drift term. We use it also to study price distributions under probability measures different to the ones in which they are simulated (see also Chap. 13).

B.2 Replication of Contingent Claims: Martingale Representation

Consider an economy that puts at our disposal a number of assets $\mathbf{S}_t = (S_t^{(1)}, \ldots, S_t^{(n)})$, so that $S_t^{(i)}$ is the time-t price of the i'th asset. There is a market for trading these assets. Thus, at any time t, a market participant, of wealth V_t say, will have a proportion of wealth allocated to a portfolio $\boldsymbol{\theta}_t = (\theta_t^{(1)}, \ldots, \theta_t^{(n)})$, with the remainder held in some deposit account, so that

$$V_t = \varphi_t B_t + \boldsymbol{\theta}_t \cdot \mathbf{S}_t, \tag{B.8}$$

where B_t is the value at t of one unit invested in the deposit account at time zero, and $\varphi_t B_t$ is the wealth not invested in $\mathbf{S}$. Because any value kept in the deposit account grows at some positive rate, it is more useful to express asset prices in terms of B, writing $\tilde{V}_t \equiv B_t^{-1} V_t$, $\tilde{\mathbf{S}}_t \equiv B_t^{-1} \mathbf{S}_t$. The wealth equation (B.8) then becomes

$$\tilde{V}_t = \varphi_t + \boldsymbol{\theta}_t \cdot \tilde{\mathbf{S}}_t, \tag{B.9}$$

so that, as we expect, in any time interval where the holdings φ and $\boldsymbol{\theta}$ are kept constant, the growth in discounted wealth $\tilde{V}$ derives only from growth in the discounted assets $\tilde{\mathbf{S}}$.

Of course, funds *may* be switched between the holdings in $\mathbf{S}$ and the deposit account, but it is natural to suppose that no *new* wealth can be injected, in which case the portfolio of holdings $(\varphi, \boldsymbol{\theta})$ is said to be *self-financing*. The consequence of V being self-financing is then that

$$\tilde{V}_t = \tilde{V}_0 + \int_0^t \boldsymbol{\theta}_u \cdot d\tilde{\mathbf{S}}_u, \tag{B.10}$$

so that the *discounted wealth is the integral of the portfolio holdings against the discounted asset price process.*

The fundamental theorem of asset pricing, formalised by Harrison & Kreps [57] and Harrison & Pliska [58], and formulated in more general setting in the work of Delbaen & Schachermayer (for example, [34] and [35]), states that arbitrage is excluded if and only if there is some *equivalent martingale measure* under which discounted asset price processes are martingales. This implies that the price of a contingent claim can be computed as the expectation in the martingale measure of the discounted payoff of that claim. If the market is also *complete*, so that all claims can be replicated perfectly,[1] then the martingale measure (and hence the market price for any claim) is unique.

Now if $\tilde{\mathbb{P}}$ is a measure under which $\tilde{\mathbf{S}}$ is a martingale, and $Y = f(\tilde{\mathbf{S}}_T)$ is a contingent claim on $\tilde{\mathbf{S}}$, the discounted time-t price of Y, $\tilde{\pi}_{t,T}$, say, being the price of a traded asset, is itself a $\tilde{\mathbb{P}}$ martingale. It follows that $\tilde{\pi}$ has a representation as

$$\tilde{\pi}_{t,T} = B_t^{-1} \pi_{t,T} = \tilde{\mathbb{E}}[B_T^{-1} Y | \mathscr{F}_t]. \tag{B.11}$$

In the absence of any other condition enforcing a unique price for the claim Y, there will be potentially as many prices $\tilde{\pi}$ for Y as there are market agents, each price reflecting that agent's own risk aversion. If the market is *complete*, however, there *is* a price-enforcing mechanism: the price of Y will be the cost V_0^Y of setting up a portfolio worth

$$V^Y(0) = \varphi_0^Y + \boldsymbol{\theta}_0 \cdot \mathbf{S}_0 \tag{B.12}$$

at time zero and

$$V^Y(T) = Y \tag{B.13}$$

at time T.

The existence of a unique process θ^Y that makes the wealth equation (B.10) true is a consequence of the martingale property of the price processes $\tilde{\pi}_{t,T} = \tilde{V}^Y(t)$ and $\mathbf{S}_t$ and the *martingale representation theorem* (see Rogers & Williams [94]).

Theorem 2 *Let X be a local martingale on the filtered probability space $(\Omega, \mathscr{F}, (\mathscr{F}_t), \mathbb{P})$, and assume that $(\mathscr{F}_t)$ is the filtration generated by X. Then, any local martingale M adapted to $(\mathscr{F}_t)$ has a representation as*

$$M_t = M_0 + \int_0^t H_u dX_u \tag{B.14}$$

where H is previsible with respect to $(\mathscr{F}_t)$. Moreover, H is unique up to sets of measure zero.

Because the claim price process $\tilde{\pi}_{t,T}$ and the asset price process $\mathbf{S}$ are both $\tilde{\mathbb{P}}$-martingales, the martingale representation theorem shows the existence of a strategy with which to *hedge* the claim Y by trading in the assets $\mathbf{S}$.

---

[1] By this we mean that for every time-T claim Y one can find a portfolio $\tilde{V}_t^Y = \tilde{V}_0 + \int_0^t \boldsymbol{\theta}_u \cdot d\tilde{\mathbf{S}}_u$ such that $V_T = Y$.

B.3 Change of Numeraire

In writing the wealth equation (B.10) we defined $\tilde{\mathbf{S}}_t \equiv B_t^{-1}\mathbf{S}_t$ and $\tilde{V} \equiv B_t^{-1}V_t$ by expressing the prices of assets and the wealth V in units of the deposit account. One says that the deposit account is being used as *numeraire*.

There is nothing that keeps us from using as numeraire the value of a different asset, and in fact changing numeraire is a powerful modelling and computational technique. Geman and Jamshidian were the first to employ this idea. Suppose X is the price of any traded asset (scaled by its time-zero value); for reasons that will soon become obvious, we need to assume $X_t > 0$ for each t. Then, because by definition of $\tilde{\mathbb{P}}$ all discounted assets are $\tilde{\mathbb{P}}$-martingales, we have that

$$\frac{X_t}{B_t} \text{ is a } \tilde{\mathbb{P}}\text{-martingale.} \tag{B.15}$$

This allows us to define a new measure, $\mathbb{P}^X$ say, whose Radon-Nikodym derivative is given for every t by

$$\zeta_t = \frac{X_t}{B_t} = \left.\frac{d\mathbb{P}^X}{d\tilde{\mathbb{P}}}\right|_{\mathscr{F}_t}. \tag{B.16}$$

It then follows, for any given process M, that

$$B_t^{-1}M_t \text{ is a } \tilde{\mathbb{P}}\text{-martingale} \iff X_t^{-1}M_t \text{ is a } \mathbb{P}^X\text{-martingale,} \tag{B.17}$$

so for any claim Y maturing at T we can write the equivalent expressions

$$\pi_{t,T} = \tilde{\mathbb{E}}\left[\frac{B^{-1}(T)}{B^{-1}(t)}Y \,\middle|\, \mathscr{F}_t\right] = \mathbb{E}^X\left[\frac{X^{-1}(T)}{X^{-1}(t)}Y \,\middle|\, \mathscr{F}_t\right], \tag{B.18}$$

where the first expectation happens under $\tilde{\mathbb{P}}$ and the second under $\mathbb{P}^X$. For example, if one takes for X the price process of the bond maturing at time T, the price of any claim Y received at T is

$$\pi_{t,T} = \tilde{\mathbb{E}}\left[\frac{B^{-1}(T)}{B^{-1}(t)}Y \,\middle|\, \mathscr{F}_t\right] = D_{t,T}\mathbb{E}^T\left[Y \,\middle|\, \mathscr{F}_t\right], \tag{B.19}$$

where $D_{t,T}$ is the observed time-t price of the T-bond, so that $D_{T,T} \equiv 1$, and where the expectation is now in the T-forward measure in which asset prices discounted by the T-bond are martingales. The price of Y can now be computed as the expectation of Y in the T-forward measure.

An in-depth account of martingale theory and stochastic processes, which we have used here, is Rogers & Williams [94]. Our description of self-financing portfolios closely follows the article of Rogers [92], which shows how ideas of economic equilibrium lead directly to the existence of equivalent martingale measures.

References

1. Andersen, L. (2007). *Efficient simulation of the Heston stochastic volatility model*. Bank of America.
2. Andersen, L., & Broadie, M. (2004). A primal-dual simulation algorithm for pricing multi-dimensional American options. *Management Science, 50*(9), 1222–1234.
3. Andersen, L. B. G., & Piterbarg, V. V. (2006). Moment explosions in stochastic volatility models. *Finance and Stochastics*.
4. Andreasen, J. (2007). A lego approach to hybrid modeling. In *WBS 4th fixed income conference*, London.
5. Andreasen, J. (2007). Closed form pricing of FX options under stochastic rates and volatility. *International Journal of Theoretical and Applied Finance, 11*(3), 277–294.
6. Artzner, P., Delbaen, F., Eber, J. M., & Heath, D. (1997). Thinking coherently. *Risk, 10*(11), 68–71.
7. Artzner, P., Delbaen, F., Eber, J. M., & Heath, D. (1999). Coherent measures of risk. *Mathematical Finance, 9*(3), 203–228.
8. Arvanitis, A., & Gregory, J. (2001). *Credit: The complete guide to pricing, hedging and risk management*. Risk Books.
9. Barndorff-Nielsen, O. E., Mikosch, T., & Resnick, S. I. (2001). *Lévy processes: Theory and applications*. Birkhäuser.
10. Baxter, M., & Rennie, A. (1998). *Financial calculus: An introduction to derivative pricing*. Cambridge University Press.
11. Benhamou, E., Rivoira, A., & Gruz, A. (2008). Stochastic interest rates for local volatility hybrid models. Available at SSRN: http://ssrn.com/abstract=1107711.
12. Bennani, N., & Dahan, D. (2005). *An extended market model for credit derivatives*. Société Générale.
13. Bertoin, J. (1998). *Cambridge tracts in mathematics. Lévy processes*. Cambridge University Press.
14. Bielecki, T. R., & Rutkowski, M. (2004). *Credit risk: Modeling, valuation, and hedging*. Springer.
15. Black, F., & Scholes, M. (1973). The pricing of options and corporate liabilities. *Journal of Political Economy, 81*, 637–654.
16. Boyarchenko, S. I., & Levendorskii, S. Z. (2002). *Non-Gaussian Merton-Black-Scholes theory*. Singapore: World Scientific.
17. Brace, A., Gatarek, D., & Musiela, M. (1997). The market model of interest rate dynamics. *Mathematical Finance, 7*.
18. Brigo, M., & Mercurio, F. (2006). *Interest rate models—theory and practice*. Springer.
19. Broadie, M., & Cao, M. (2008). Improved lower and upper bound algorithms for pricing American options by simulation. *Quantative Finance, 8*(8), 845–861.

G. Cesari et al., *Modelling, Pricing, and Hedging Counterparty Credit Exposure,*
Springer Finance, DOI 10.1007/978-3-642-04454-0,
© Springer-Verlag Berlin Heidelberg 2009

20. Broadie, M., & Glasserman, P. (2004). A stochastic mesh method for pricing high-dimensional American options. *Journal of Computational Finance, 7*(4), 35–72.

21. Cairns, A. (2004). *Interest rate models: An introduction.* Princeton University Press.

22. Canabarro, E., & Duffie, D. (2003). *Measuring and marking counterparty risk: Asset/liability management of financial institutions.* Euromoney Books.

23. Canabarro, E., Picoult, E., & Wilde, T. (2003). Analysing counterparty risk. *Risk, 16*(9), 117–122.

24. Carr, P., & Madan, D. (1998). Option valuation using the fast Fourier transform. *Journal of Computational Finance, 2*, 61–73.

25. Carr, P., & Wu, L. (2004). Time-changed Lévy processes and option pricing. *Journal of Financial Economics, 71*(1), 113–141.

26. Carr, P., Geman, H., Madan, D., & Yor, M. (2002). The fine structure of asset returns: An empirical investigation. *The Journal of Business, 75*(2), 305–333.

27. Carr, P., Geman, H., Madan, D., & Yor, M. (2003). Stochastic volatility for Lévy processes. *Mathematical Finance, 13*(3), 345–382.

28. Constantinides, G. M. (1992). A theory of the nominal term structure of interest rates. *The Review of Financial Studies, 5*, 531–522.

29. Cont, R. (2001). Empirical properties of asset returns: Stylized facts and statistical issues. *Quantitative Finance, 1*(2).

30. Cont, R., & Tankov, P. (2004). *Chapman & Hall CRC financial mathematics series. Financial modelling with jump processes.*

31. Cox, J. (1975). *Notes on option pricing I: Constant elasticity of variance diffusions* (Working Paper). Stanford University. Reprinted in *Journal of Portfolio Management (1996), 22*, 1517.

32. Cox, J., & Ross, S. (1976). The valuation of options for alternative stochastic processes. *Journal of Financial Economics, 3*(1–2), 145–166.

33. Delbaen, F. (2000). *Coherent risk measures.* Cattedra Galileiana, Scuola Normale Superiore di Pisa.

34. Delbaen, F., & Schachermayer, W. (1994). A general version of the fundamental theorem of asset pricing. *Mathematische Annalen, 123*, 463–520.

35. Delbaen, F., & Schachermayer, W. (1997). *The fundamental theorem of asset pricing for unbounded stochastic processes.* Preprint.

36. Derman, E. (1995). The local volatility surface. *Goldman Sachs quantitative strategies research notes.*

37. Derman, E., & Kani, I. (1994). Riding on a smile. *Risk, 7*, 32–39.

38. Duffie, D., & Huang, M. (1996). Swap rates and credit quality. *Journal of Finance, 6*, 79–406.

39. Duffie, D., & Singleton, K. J. (2003). *Princeton series in finance. Credit risk: Pricing, measurement, and management.*

40. Duffie, D., Pan, J., & Singleton, K. (2000). Transform analysis and asset pricing for affine jump-diffusions. *Econometrica, 68*(6), 1343–1376.

41. Dupire, B. (1994). Pricing with a smile. *Risk.*

42. Eberlein, E., Keller, U., & Prause, K. (1998). New insights into smile, mispricing, and value at risk: The hyperbolic model. *The Journal of Business, 71*(3), 371–405.

43. Evans, E. (2003). *Domain-driven design: Tackling complexity in the heart of software.* Addison-Wesley.

44. Filipovic, D. (2009). *Term-structure models—A graduate course.* Springer.

45. Flesaker, B., & Hughston, L. P. (1996). Positive interest. *Risk, 9*(1), 46–49.

46. Fries, P. C. (2005). Foresight bias and suboptimality correction in Monte-Carlo pricing of options with early exercise: Classifcation, calculation & removal. http://www.christian-fries.de/finmath/foresightbias.

47. Gamma, H., Helm, R., Johnson, R., & Vissides, J. (1997). *Design patterns, elements of reusable object-oriented software.* Addison-Wesley.

48. Gatheral, J. (2007). *Case studies in financial modelling.* Course Notes.

49. Geman, H., Madan, D., & Yor, M. (2000). *Quantitative analysis in financial markets. Asset prices are Brownian motion: Only in business time.* World Scientific Publishing Company.

50. Glassermann, P. (2004). *Monte Carlo Methods in financial engineering*. Springer.
51. Gorton, I. (2006). *Essential software architecture*. Springer.
52. Gregory, J. (2009). Being two-faced over counterparty credit risk. *Risk, 2*, 86–90.
53. Grune, D., Bal, H., Jacobs, C., & Langendoen, K. (2000). *Modern compiler design*. Wiley.
54. Gyöngy, I. (1986). Mimicking the one-dimensional distributions of processes having an Ito differential. *Probability Theory and Related Fields, 71*, 501–516.
55. Hagan, P., Kumar, D., Lesniewski, A. S., & Woodward, D. E. (2002). Managing smile risk. *Wilmott Magazine*, 84–108.
56. Harrell, F. E., & Davis, C. E. (1982). A new distribution-free quantile estimator. *Biometrika, 69*(3), 635–640.
57. Harrison, M., & Kreps, D. (1979). Martingales and arbitrage in multiperiod securities markets. *Journal of Economic Theory, 20*, 381–408.
58. Harrison, M., & Pliska, S. (1981). Martingales and stochastic integrals in the theory of continuous trading. *Stochastic Processes and their Applications, 11*, 313–316.
59. Haugh, M. B., & Kogan, L. (2004). Pricing American options: A duality approach. *Operations Research, 52*(2), 258–270.
60. Heath, D., Jarrow, R., & Morton, A. (1992). Bond pricing and the term structure of interest rates: A new methodology for contingent claims valuation. *Econometrica, 60*, 77–105.
61. Heston, S. (1993). A closed-form solution for options with stochastic volatility. *Review of Financial Studies, 6*, 327–343.
62. Hull, J. (2003). *Options, futures, and other derivatives*. Prentice-Hall.
63. Hull, J., & White, A. (1990). Pricing interest-rate derivative securities. *Journal of Financial and Quantitative Analysis, 3*(4), 573–592.
64. Hunt, P. J., & Kennedy, J. E. (2004). *Financial derivatives in theory and practice*. Wiley.
65. Hyer, T. (1996). *Impact of bundling parameters*. Bankers Trust.
66. Hyer, T. (1996). *Extensions to the bundling decider*. Bankers Trust.
67. Kahl, C., & Jaeckel, P. (2005). Not-so-complex logarithms in the Heston model. *Wilmott Magazine*, 74–103.
68. Karatzas, I., & Shreve, S. E. (1991). *Brownian motion and stochastic calculus*. Springer.
69. Kecman, M. (2009). *An analysis of the least-squares Monte-Carlo pricing approach*. MS dissertation, Imperial College, London.
70. Kloeden, P. E., & Platen, E. (1999). *Numerical solution of stochastic differential equations*. Springer.
71. Kou, S. G. (2002). A jump-diffusion model for option pricing. *Management Science, 48*(8), 1086–1101.
72. Lando, D. (2004). *Princeton series in finance. Credit risk modeling: Theory and applications*.
73. Levine, J. R. (2009). *Flex & Bison*. O'Reilli.
74. Levine, J. R., Mason, T., & Brown, D. (1995). *Lex & Yacc*. O'Reilli.
75. Lomibao, D., & Zhu, S. (2005). *Capital markets risk management. A conditional approach for path-dependent instruments*. Bank of America, N.A., New York.
76. Longstaff, F., & Schwartz, E. (2001). Valuing American options by simulation. A simple least-squares approach. *Review of Financial Studies, 14*, 113–147.
77. Madan, D. B., Konikov, M., & Marinescu, M. (2004). Credit and basket default swaps. *Journal of Credit Risk, 2*(1).
78. McNeil, A. J., Frey, R., & Embrechts, P. (2005). *Princeton series in finance. Quantitative risk management: Concepts, techniques, and tools*.
79. Mendoza, R., Carr, P., & Linetsky, V. Time changed Markov processes in unified credit-equity modelling. *Math Finance*, forthcoming.
80. Merton, R. C. (1974). On the pricing of corporate debt: The risk structure of interest rates. *Journal of Finance, 29*(2), 449–470.
81. Merton, R. (1976). Option pricing when underlying stock returns are discontinuous. *Journal of Financial Economics, 3*(1–2), 125–144.
82. Milstein, G. N. (1995). *Numerical integration of stochastic differential equations*. Springer.

83. Øksendal, B. (1992). *Stochastic differential equations*. Springer.
84. Parr, T. (2007). *The definitive ANTLR reference: Building domain-specific languages*. The Pragmatic Bookshelf.
85. Pelsser, A. (2000). *Efficient methods for valuing interest rate derivatives*. Springer.
86. Protter, P. E. (2004). *Stochastic integration and differential equations* (2nd ed.). Springer.
87. Pykhtin, M. (2005). *Counterparty credit risk modelling*. Risk Books.
88. Pykhtin, M., & Zhu, S. (2006). Measuring counterparty credit risk for trading products under base II. In M. Ong (Ed.), *Basel II handbook*. Risk Books.
89. Rebonato, R. (2002). *Modern pricing of interest-rate derivatives—The libor market model and beyond*. Princeton University Press.
90. Rogers, L. C. G. (1997). The potential approach to the term structure of interest rates and exchange rates. *Mathematical Finance, 7*, 157–164.
91. Rogers, L. C. G. (2002). Monte Carlo valuation of American options. *Mathematical Finance, 12*, 271–286.
92. Rogers, L. C. G. (2005). *The origins of risk-neutral pricing and the Black-Scholes formula*. Preprint, University of Bath.
93. Rogers, L. C. G., & Williams, D. (2000). *Diffusions, Markov processes and martingales: Vol. 1. Foundations*. Wiley.
94. Rogers, L. C. G., & Williams, D. (2000). *Diffusions, Markov processes and martingales: Vol. 2. Itô calculus*. Wiley.
95. Schönbucher, P. (2003). *Credit derivatives pricing models: Model, pricing and implementation*. Wiley Finance.
96. Schoutens, W., & Symens, S. (2003). The pricing of exotic options by Monte-Carlo simulations in a Lévy market with stochastic volatility. *International Journal for Theoretical and Applied Finance, 6*(8), 839–864.
97. Schroder, M. (1989). Computing the constant elasticity of variance option pricing formula. *Journal of Finance, 44*, 211–219.
98. Shreve, S. E. (2008). *Stochastic calculus for finance II: Continuous-time models*. Springer Science.
99. Sidenius, J., Piterbarg, V. V., & Andersen, L. B. G. (2008). A new framework for dynamic credit portfolio loss modeling. *International Journal of Theoretical and Applied Finance, 11*(2), 163–197.
100. Sorensen, E., & Bollier, T. (1994). Pricing swap default risk. *Financial Analysts Journal, 50*, 23–33.
101. Tang, Y., & Li, B. (2007). *Quantitative analysis, derivatives modeling, and trading strategies in the presence of counterparty credit risk for fixed-income market*. World Scientific Publishing Co. Pte. Ltd.
102. Tasche, D. (1999). *Risk contributions and performance measurement*. Preprint, Department of Mathematics, TU-München.
103. Tilley, J. A. (1993). Valuing American options in a path simulation model. *Transactions, Society of Actuaries Schaumburg, 45*, 499–520.
104. Unified Modeling Language. www.uml.org.
105. Vasicek, O. A. (1987). Probability of loss on loan portfolio, Moody's.
106. Williams, D. (1991). *Probability with martingales*. Cambridge University Press.

Index

Printed by Books on Demand, Germany